VEGETATION

Edited by **Allan Sebata**

Vegetation
http://dx.doi.org/10.5772/intechopen.69411
Edited by Allan Sebata

Contributors
Fabiana Firetti, Manuel Pulido Fernández, J. Francisco Lavado Contador, Susanne Schnabel, Álvaro Gómez Gutiérrez, Javier Lozano-Parra, Miomir Jovanović, Bojan Vračarević, Miško Milanović, Albert Chirima, Peter Mundy, Andre F VanRooyen, Nicholas Ncube, Eusebio Cano Carmona, Carmen Bartolomé, Musarella Carmelo María, Piñar Fuentes J.C., Quinto Canas Ricardo, Pinto Gomes C.J., Cano Ortiz Ana, Cristiane Sarmento, Paramesha Mallegowda, Ganesan Rengaian, Siddappa Setty

Published by InTech
Janeza Trdine 9, 51000 Rijeka, Croatia

Publishing Process Manager Marina Dusevic
Technical Editor SPi Global
Cover InTech Design team

First published March, 2018
Printed in Croatia
Legal deposit, Croatia: National and University Library in Zagreb

Additional hard copies can be obtained from orders@intechopen.com

Vegetation, Edited by Allan Sebata
p. cm.
Print ISBN 978-953-51-3829-7
Online ISBN 978-953-51-3830-3

For EU product safety concerns:
IN TECH d.o.o., Prolaz Marije Krucifikse Kozulić 3, 51000 Rijeka, Croatia,
info@intechopen.com or visit our website at intechopen.com.

Meet the editor

Dr. Allan Sebata is a senior lecturer at the Department of Forest Resources and Wildlife Management at the National University of Science and Technology, Zimbabwe. He received his PhD degree in Rangeland Ecology from the same university in 2012. Dr. Allan Sebata earned his BSc Honours in Agriculture (Animal Science) and MSc degree in Animal Science from the University of Zimbabwe. He lectures both undergraduate and postgraduate courses in ecology. In addition, currently, he is supervising undergraduate and postgraduate research projects. His research in recent years has focused on herbivory and management of rangelands for both livestock and wild ungulate uses. Dr. Sebata has published research papers in international journals while also being actively involved in peer review of manuscripts.

Contents

Preface

Vegetation is an assemblage of plant species, without specific reference to particular taxa, life forms, structure or spatial extent. Different vegetation types constitute various plant communities such as forests, savannas and grasslands. Vegetation, particularly forests, contributes to carbon storage and water cycle regulation and acts as habitats for birds and animals. However, anthropogenic activities such as logging and overgrazing result in vegetation losses, which negatively affect the provision of ecosystem goods and services. Approximately 12 million km^2 of forests and woodlands have been removed globally since 1700 AD. To mitigate vegetation losses, particularly in forests, land cover changes need to be quantified and monitored regularly.

This book looks at land cover changes across the world in Africa, Asia, Europe and South America. Normalized difference vegetation index (NDVI) and CORINE Land Cover are used to monitor vegetation changes in southern Serbia. Overtime, many dense forests around the world have been converted into agricultural land, grazing areas or settlements through intense human activities. Vegetation changes that took place in Spain and Portugal over a 60-year period as a result of progressive land clearance to make way for pastures are documented.

Land cover changes due to anthropogenic activities in a Miombo woodland in Africa are studied over a 27-year period using geographical information systems and remote sensing. It is estimated that neotropical forests hold almost half of the carbon pool in tropical forests in addition to being key contributors to global biodiversity, warranting detailed studies of this important ecosystem. In this book, the effect and extent of anthropogenic factors in the loss of neotropical forests as well as conservation efforts to restore degraded portions are examined.

Climate is a key determinant of island forest vegetation with endemic species being predominant as they are adapted to the unique local conditions. Vegetation of the Spanish island of Hispaniola (Dominican Republic) is divided into two major groups, viz., dry and cloud forests, each with different vegetation subtypes, structure and diversity, and studied in detail. The restoration of a degraded wildlife corridor through re-establishing the original native vegetation in India is also reported.

Apomixis is a form of asexual reproduction via seed and is considered as an alternative reproductive strategy in many plant families. It involves development of a new organism from an unreduced and unfertilized egg cell and is considered a dead end of evolution due to lack of genetic variability. The occurrence of apomixes in several plant families in the neotropical region is explored.

Dr. Allan Sebata
Department of Forest Resources and Wildlife Management
National University of Science and Technology (NUST), Bulawayo, Zimbabwe

Chapter 1

Comparing NDVI and Corine Land Cover as Tools for Improving National Forest Inventory Updates and Preventing Illegal Logging in Serbia

Jovanović M. Miomir, Miško M. Milanović and
Bojan R. Vračarević

Additional information is available at the end of the chapter

http://dx.doi.org/10.5772/intechopen.71845

Abstract

National forest inventories (NFIs) in Serbia have been carried out very rarely (every 20 years), while the last two official estimates of forest areas (for 2011 and 2014) are very imprecise, because they are based on the cadastral data (and Serbia is well known for the lack of cadastre updating). Although forest conservation policymakers in Serbia still have limited financial, human, and political resources, over the past two decades, publicly available, remotely sensed satellite data on deforestation and degradation have dramatically reduced evaluation costs. Since municipalities in Southern Serbia experienced a 15% loss of forest area in the 2006–2014 period, as the obvious result of forceful, rapid process of illegal logging, this study evaluates the possible use of two remote sensing techniques: normalized difference vegetation index (NDVI) and CORINE land cover (CLC) databases for preventing illegal logging in Serbia. It clearly shows that NDVI is very promising for Serbia and also for other post-socialist countries that very rarely carry out national forest inventories (NFIs), and where unrecorded, illegal logging can exceed the legal harvest by a factor of 10.

Keywords: CORINE land cover, illegal logging, national forest inventory, normalized difference vegetation index, Serbia

1. Introduction

Like many other developing and post-socialist countries where the basic information on the current state of forests and other ecosystems is often inadequate, fragmentary, or outdated [1], in Serbia national forest inventories (NFIs) have been carried out very rarely, at roughly

20-year intervals: in 1961, 1979, and 2003–2006. Since 2006, official estimates of forest areas have been made only for 2011 and 2014, but they are very imprecise, since they are based on the cadastral data (and Serbia is well known for the lack of cadastre updating) [2, 3].

While Serbian state-owned forests (48% of the nation's forest resources) are managed mainly by the state forest enterprises, according to the management plans prepared on the 10-year basis, the basic information on the current state of private forests is in a much worst condition. Previous forest census completed in 1979 covered only state forests and national parks, without even including private forests (new inventory completed in 2006, finally included private forests). Private forests constitute 52% of the nation's forest resources, and are characterized by very small plots (average size: 0.3 ha). Nevertheless, Serbia adopted following official definition of forest area: "The forest includes all the inventory unit areas larger than 0.5 ha covered with forest trees…" [2], which is very similar to FAO definition [4]. Actually, NFI features for individual countries have been developed over time to accommodate their unique topographies, climates, forest types, commercial interests, etc. [5–7]. Although FAO's reference definition is used as the basis for the national definitions of forest by many countries, national definitions can vary considerably: minimum forest area in Czech Republic is 0.04 ha, Austria (0.05 ha), China (0.0667 ha), Germany (0.1 ha), Estonia (0.1 ha), Ireland (0.1 ha), Latvia (0.1 ha), Lithuania (0.1 ha), Luxembourg (0.1 ha), Slovenia (0.25 ha), Slovak Republic (0.3 ha), USA (0.4 ha), etc. [8]. Unfortunately, 0.5 ha does not fit Serbia well, since privately owned forest parcels, which account for half of the total forest area of the country, cover much smaller areas—average private holding is only 0.3 ha [9].

Since illegal logging contributes up to 30% of the global market, in excess of US $20 billion a year, we cannot rely on official production statistics to capture deforestation. For example, even the latest forest resource data from official statistics in Central and Eastern European countries often do not consider forest degradation and illegal logging [10]. World Bank estimates that unrecorded, illegal logging in some of Central and Eastern European countries, like Albania, exceeded the legal harvest by a factor of 10 [3]. Also, in Serbia there are no unique records about illegal activities in forestry [11]. Illegal logging in Serbia is most intense exactly in the areas adjacent to the territories of Kosovo (Kosovo is ranked as the one of the worst illegal logging offenders in the world with Indonesia [12, 13]) which are formally under Serbian forest estates Vranje, Kursumlija, Leskovac, Raska, and Leposavic), territories to which Serbian authorities have limited access [3, 14, 15].

In order to prevent illegal logging, it is obviously most relevant to concentrate on areas where the illegal logging is most intense, and it is exactly in the southern Serbian municipalities, like Kursumlija, that are adjacent to the territories of Kosovo.

Although forest conservation policymakers in the most post-socialist European countries still have limited financial, human, and political resources, over the past two decades, publicly available, remotely sensed satellite data on deforestation and degradation have dramatically reduced evaluation costs [16, 17]. Such an approach could be much cheaper and the series of data would be quite useful for monitoring the forest cover [12].

Remote sensing is the detection, recognition, or evaluation of objects by means of distant sensing or recording devices [18]. NDVI and CLC belong to these very promising remote sensing techniques that allow monitoring forest changes over time [19, 20]. One of the main differences between NDVI and CLC is that when NDVI focuses on the vegetation cover and its status, CLC has a much broader scope and distinguishes agricultural areas, forests and semi-natural areas, artificial surfaces, urban fabric, industrial, commercial, and transport units, bodies of water, wetlands, glaciers and perpetual snow, and other features [21]. Normalized difference vegetation index (NDVI) [22] is one of the most widely used vegetation indices (VIs) which focuses on the vegetation cover and its status [21–28]. NDVI, like all VIs, relates the spectral absorption of chlorophyll in the red with a reflection phenomenon in the near-infrared, influenced by the leaf structure type [29]. NDVI also has the advantage of allowing comparisons between images acquired at different times [30]. On the other hand, CLC is a European program launched in 1985 by the European Commission, aimed at obtaining a comparable dataset of land cover for Europe. The aim of CLC is to gather information related to the environment on certain priority topics for the European Union: air, water, soil, land cover, coastal erosion, biotopes, etc. The main goal of the CLC program is to unify heterogeneous thematic cartographies of Europe at various levels (international, national, regional, local), and to update data at regular intervals, every 5–10 years [31, 32]. CLC is a map of the European environmental landscape based on interpretation of satellite images. The data have been validated using local cartography and ground surveys [33–35]. CLC also has an NDVI module for creating vegetation maps.

Advances in remotely sensed data availability have, obviously, created significant new opportunities to map changes in land cover and forest patterns and enhance understanding of the effectiveness and efficiency of forest conservation policy in Serbia [36–38].

2. Materials and methods

In order to evaluate the possible use of Corine land cover (CLC) and NDVI for preventing abrupt illegal logging in southern Serbian municipalities [39], our study was carried out for the municipality of Topola, located in central Serbia, and the Municipality of Kursumlija that lies in southern Serbia, bordering the territories of Kosovo (**Figure 1**).

NDVI data for both municipalities are based on Landsat 5 Thematic Mapper (TM) satellite images for 2006, 2011, 2012 and 2014, which were created during spring/summer (August), with minimum clouds (10–20%; [40]).

NDVI was used and necessary corrections/transformations were applied for visible red in constellation with the infrared spectrum of satellite images using the following procedure: GIS Analysis/Mathematical Operation/Image Calculator, and then the equation NDVI = (NIR − RED)/(NIR + RED), in which NIR is the near-infrared channel and RED is the red channel from the visible part of the spectrum [41, 42]. Basic tasks included analysis and photo interpretation of elements, occurrences, and processes detected on images

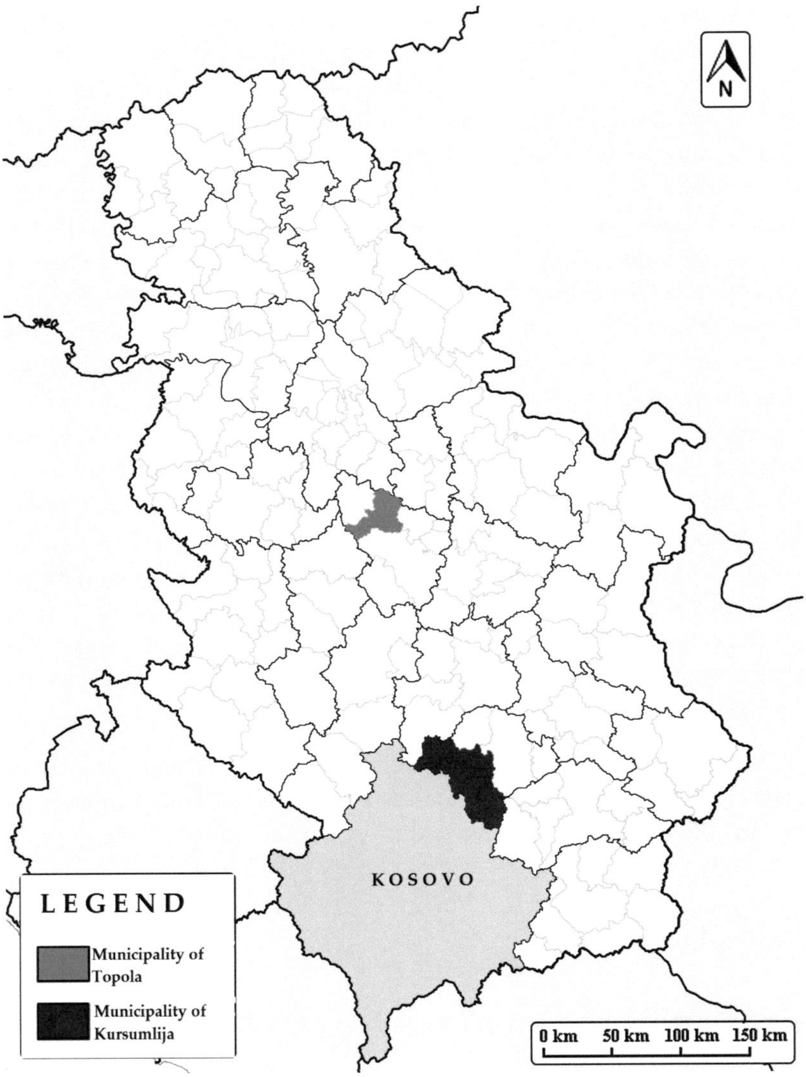

Figure 1. Locations of the municipalities of Kursumlija and Topola.

using specialized GIS software (Idrisi 15-Andes) for processing remotely sensed images through application of NDVI. Since shadow areas were less than 5% in the Municipality of Kursumlija and less than 3% in the Municipality of Topola, no topographic corrections were made (**Table 1**).

Prior to change detection, a normalized difference vegetation index (NDVI) images were generated and a threshold classification technique was applied. Using the NDVI, threshold technique involves less time and data, but selecting an accurate NDVI threshold can be difficult and still remains a challenge. For example, some surveys show that soils have a highly variable NDVI and the mean value (0.20–0.21) that is much larger than the NDVI

LANDSAT 5 (TM sensor)	Wavelength (pm)	Resolution (m)
Band 1	0.45–0.52	30
Band 2	0.52–0.60	30
Band 3	0.63–0.69	30
Band 4	0.76–0.90	30
Band 5	1.55–1.75	30

Table 1. Different spectral channels of images (produced by LANDSAT satellite) used in this paper.

commonly used (< 0.05). This problem is most severe in areas with sparse vegetation cover (e.g. grassland and shrubland) where typical seasonal NDVI values are in the sensitive range (0.2 < NDVI pixel < 0.4) [26]. Also, although the most common minimum NDVI value for broad-leaved forests is 0.4, even the 0.25 NDVI threshold was appropriate for some deciduous forest types (for northeastern China) [43] and also for temperate forests [44]. Previously, AVHRR-derived NDVI data advanced very high resolution radiometers (AVHRR) have been used to define the length of temperate forest growing seasons, using NDVI threshold values ranging from 0.25 [44] to 0.45 [45]. One survey of Northern Ghana suggests that appropriate NDVI value for forests is 0.32–0.4 [46], while study of Bangladesh it is 0.25 [47]. Also, Xiao et al. suggest that the 0.25 NDVI threshold is appropriate for the onset of greenness development of deciduous forest types in northeastern China [43]. Any generalization, as well as creation of the class itself, is often an arbitrary process. Also, threshold parameters, produce arbitrary and artificial differences in values in the real world [48]. Hence, the underlying logic of the reliable use of the threshold technique should be derived experimentally by defining classes step-by-step to know if they are correct [49]. As Jansen points out, different perspectives, or so-called 'scapes', to categorization can be taken that are all equally valid and valuable [49]. The threshold used for classification here has been chosen based on using sites of known change and stability in order to define the most suitable and reliable threshold for Serbia. Extensive field surveys were then guided with a global positioning system (GPS) receiver. The data collected during these field surveys were used to determine the major types of land use in the study areas, which helped design a land cover classification and to associate the ground "truth" of a specific type of land cover with its imaging characteristics, and which helped in making a reliable threshold. In this phase, we had to use supervised classification to correct intensity, hue and saturation of pixels, and also to correct the obtained shades (that were showing vegetation), in order to precisely distinguish forests from vineyards and orchards and to produce precise distinction between shrub and broad-leaved forest for Serbia. For broad-leaved forest in Serbia, appropriate range was between 0.292 and 0.438. The NDVI images were classified into eight classes: (1) grassy areas, meadows, and pastures, (2) sparsely vegetated areas, (3) shrub vegetation, (4) vineyard, (5) orchards, (6) broad-leaved forest, (7) mixed forest, and (8) coniferous forest (**Figures 2–4**). Vegetation areas are presented with values between 0 and 1. Grassy areas, meadows, and pastures have values that range from zero up to 0.122. Sparsely vegetated areas has an NDVI value from 0.123 to 0.174, shrub vegetation between

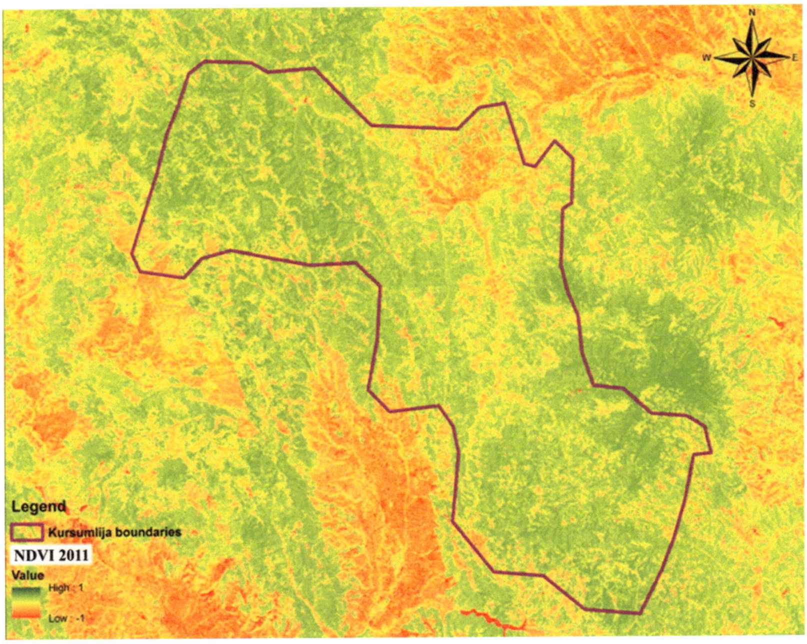

Figure 2. Vegetation cover of municipality Kursumlija for 2011 obtained by NDVI.

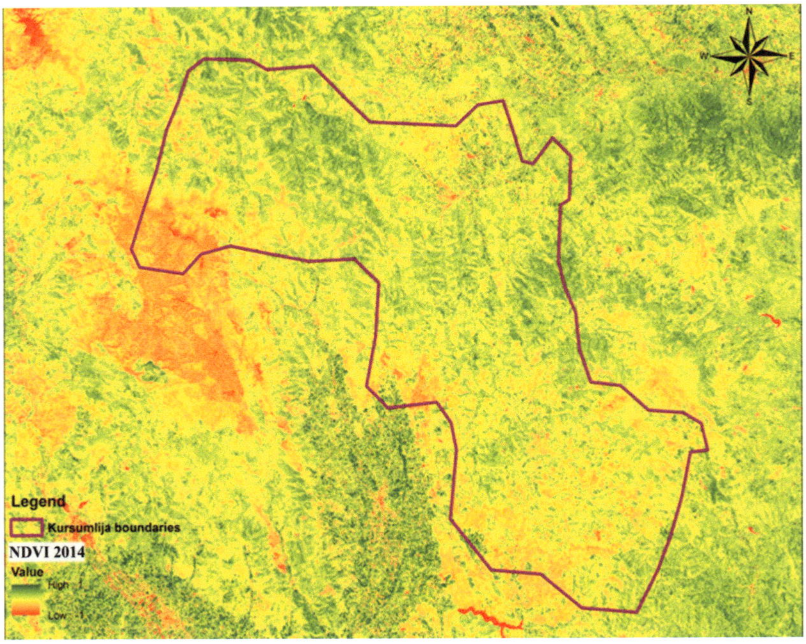

Figure 3. Vegetation cover of municipality Kursumlija for 2014 obtained by NDVI.

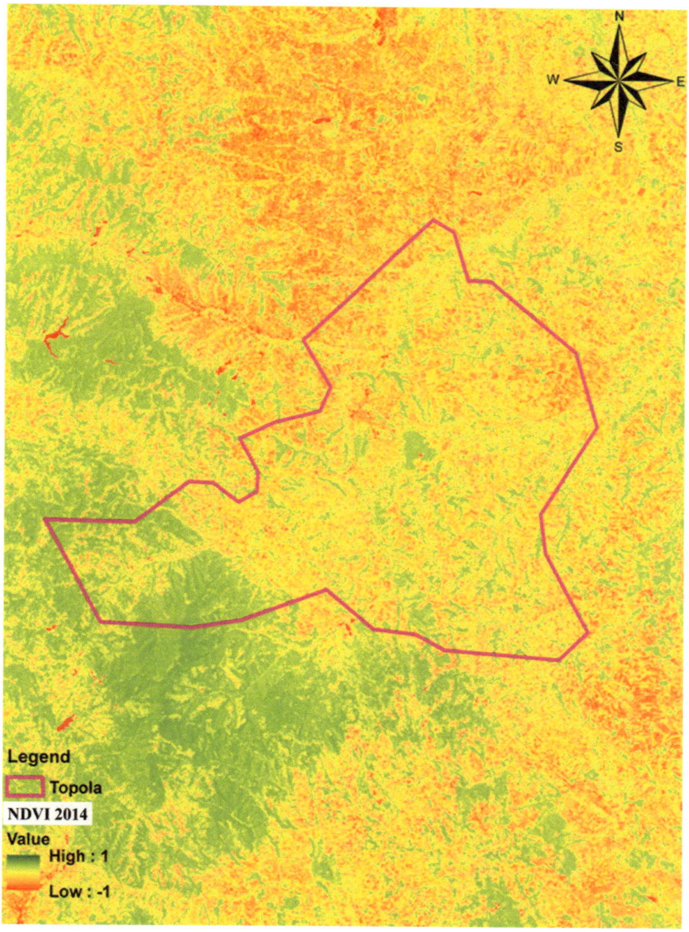

Figure 4. Vegetation cover of municipality Topola for 2014 obtained by NDVI.

0.175 and 0.230, vineyard 0.231 and 0.262, orchards 0.263 and 0.291, broad-leaved forest between 0.292 and 0.438, mixed forest between 0.439 and 0.525, and coniferous forest has an NDVI value above 0.526 [28, 50].

For creating CLC maps for the municipalities of Kursumlija and Topola, image processing was carried out and a digital elevation model was made based on the municipalities' boundaries and Landsat satellite color composites, and a pseudo-color composite with bands 4, 5, 1 and adequate contrast was applied. Datasets and maps for Serbia, mainly CLC2006 and CLC2012, were extracted from the European Environmental Agency (EEA) website, with a transfer data scale of 1:100,000 [51]. Although in the original CLC project the smallest unit is 25 ha, a recent approach yields more precise results because changes <25 and >5 ha are mapped [52]. Nevertheless, even the smallest 5 ha areas, which are highly appropriate at the EU scale, do not properly reflect the land use situation at the local scale in a country where landscapes and land use change across very short distances [53, 54]. Obviously CLC is an example of a top-down process of European standardization that follows a common system of nomenclature and focuses on common definitions and methods, and does not properly take into account unique features of individual countries of Central and Eastern Europe, like Serbia.

Finally, we performed the map accuracy assessment in order to evaluate the quality of our maps by comparing the mapped or predicted value to the observed or true condition. The basis of any accuracy assessment is a location-specific comparison of a map prediction and a ground observation. Accuracy assessment requires three primary components [55]: (1) the sampling design—choosing which locations are visited; (2) the response design, used to decide whether the map predicted and observed condition match; and (3) estimating accuracy parameters and summarizing the results of the assessment [55–60]. As Foody [56] points out, accuracy assessment has been a topic of considerable debate and research in remote sensing for many years. This is in part because the promoted standard methods such as the Kappa coefficient are not always appropriate [56]. Instead of Kappa, we have chosen different approach. The main problem with applying CLC for Serbia is that any (for CLC) unrecognizable spatial area (smaller than 4 ha) is automatically added to the closest recognizable larger area. Since CLC provided 135 of these small spatial areas (smaller than 4 ha) for Topola and 324 for Kursumlija, we concentrated exactly on them, and performed multiple GPS field surveys to compare our results with the "ground truth".

Hence, for our sampling design (1) we have chosen 64 (out of 135) of these spatial areas smaller than 4 ha (their size varied—from 0.02 to 4 ha), then (2) we compared our NDVI results with GPS field survey results, to decide whether the map predicted and observed condition match; and finally (3) we estimated accuracy parameters and summarized the results of the assessment (**Table 11**). This way not only that we could: (a) show the level of CLCs imprecision for spatial areas smaller than 4 ha (compared to the NDVI results), but also (b) we could check out the level of NDVI accuracy.

Also, since municipalities Kursumlija and Topola include mountainous areas (up to 1400 and 1800 m, respectively), and elevation, aspect, and slope are the three main topographic factors that control the distribution and patterns of vegetation in mountain areas [61–64], while among these three factors, elevation is most important [64–67], we performed statistical analysis to see whether different altitude zones have any impact on our NDVI results.

3. Results

When the CLC results for 2006 and 2012 were compared with official forest area estimates for the same years (**Tables 2–5**), some inconsistencies became apparent:

Forest areas obtained from CLC were up to 15.6% larger than the official forest area estimates. Kursumlija's forest area obtained from CLC for 2006 is 4.3% larger than the official forest area estimates for this municipality and 15.6% larger than the official forest area estimates for 2012, while Topola's forest area obtained from CLC for 2006 is 11.5% and for 2012 is 10.8% larger than the official forest area estimates.

Type of vegetation	**Kursumlija (km²)**		**Topola (km²)**	
Year	2006	2012	2006	2012
Settlements	4.60	4.87	9.11	9.30
Green urban areas	—	—	0.82	0.78
Non-irrigated arable land	0.42	0.35	36.51	36.49
Natural grasslands	25.74	25.42	14.44	14.27
Complex cultivation patterns	78.73	79.61	154.94	155.12
Land principally occupied by agriculture, with significant areas of natural vegetation	102.25	101.81	73.17	72.91
Broad-leaved forest	620.68	619.78	55.68	55.54
Coniferous forest	3.63	3.44	0.19	0.15
Mixed forest	6.14	6.01	2.12	2.00
Pastures	24.18	24.02	1.11	1.13
Transitional woodland-shrub	75.78	77.07	0.81	1.21
Sparsely vegetated areas	0.78	0.55	—	—
Total	942.93	942.93	348.9	348.9

Table 2. Land cover in the municipalities of Kursumlija and Topola obtained from CLC for 2006 and 2012.

Municipality	**Municipality total area (km²)**	**Official statistics for 2006**	**Official statistics for 2012**	**Calculated on the basis of CLC for 2006**	**Calculated on the basis of CLC for 2012**
Topola	357	52.00	52.0494	57.99	57.69
Kursumlija	952	604.41	544.2856	630.45	629.23

Table 3. Forest areas according to official statistics and calculated on the basis of CLC for the years 2006 and 2012 (in km²).

	CLC—Official statistics difference (km²)		**CLC—Official statistics difference (%)**	
Year	2006	2012	2006	2012
Topola	5.99	5.64	11.52	10.84
Kursumlija	26.04	84.94	4.31	15.61

Table 4. Difference between forest areas calculated on the basis of CLC and according to official statistics for the years 2006 and 2012 (in km² and in %).

Municipality	Municipality total area (km²)	Official statistics for 2006	Official statistics for 2011	Official statistics for 2014	Calculated on the basis of NDVI for 2006	Calculated on the basis of NDVI for 2011	Calculated on the basis of NDVI for 2014
Topola	357	52.00	52.0494	47.52	52.14	51.24	44.34
Kursumlija	952	604.41	544.2856	546.474	600.97	530.64	507.85

Table 5. Forest areas according to official statistics and calculated on the basis of NDVI for the years 2006, 2011, 2014 (in km²).

On the other hand, when compared with official forest area estimates [68–70], the NDVI results for the Municipality of Topola's forest area show a mere −0.27% difference for 2006, −1.55% for 2011, and −6.69% difference for 2014, and for the Municipality of Kursumlija - 0.57% difference for 2006, −2.51% for 2011, and −7.07% difference for 2014 (**Table 6**). For 2006 and 2011, NDVI results completely fit within the ±5% margin of error allowed for this method [71, 72], while for 2014 (−7% difference) our NDVI results exceed ±5% margin of error. Obviously, since NFI 2006, discrepancies between official forest area updates and our NDVI results are permanently growing, which opens possibilities for further analysis.

Vegetation cover in the municipalities of Kursumlija and Topola obtained from NDVI for 2006, 2011, and 2014 is presented in **Tables 7**. Since official updates for 2011 and 2014 do not contain different (coniferous, broad-leaved, and mixed) forest areas, comparison of NDVI values with official statistics was not possible.

There are huge differences between our CLC and NDVI results for forest area for Kursumlija—even 98.59 km² of a difference for 2012 (99.1%—or 97.74 km²—of these differences belong to the broad-leaved forest area category (**Tables 8–10**).

When we finally performed the map accuracy assessment, for all of our NDVI forest spatial areas, the error did not exceed ±3%. Hence, there was an extremely high degree of conformity between NDVI results and the "ground truth" (**Tables 11**).

Finally, by using SPSS software we showed completely insignificant value of Pearson's correlation coefficient (0.106) for different altitude zones (240–280, 330–450, 450–550, 750–850, and 900–1400 m) and deviation between forest areas obtained by NDVI and by GPS, clearly showing that topography does not have any relevant impact on our NDVI results.

	NDVI—Official statistics difference (km²)			NDVI—Official statistics difference (%)		
Year	2006	2011	2014	2006	2011	2014
Topola	0.14	−0.81	−3.18	0.27%	−1.55	−6.69
Kursumlija	−3.44	−13.65	−38.62	−0.57%	−2.51	−7.07

Table 6. Difference between forest areas calculated on the basis of NDVI and according to official statistics for the years 2006, 2011 and 2014 (in km² and in %).

Type of vegetation	Kursumlija (km²)			Topola (km²)		
Year	2006	2011	2014	2006	2011	2014
Coniferous forest	9.30	3.65	3.61	0.61	0.17	0.13
Mixed forest	12.42	4.95	4.72	2.08	2.02	1.95
Broad-leaved forest	579.25	522.04	499.52	49.45	49.05	42.26
Orchards	25.63	33.54	35.55	38.33	43.25	51.35
Vineyards	7.40	6.41	5.15	12.11	7.23	5.02
Shrub vegetation	23.33	15.42	13.24	12.22	4.20	3.86
Sparsely vegetated areas	5.2	4.12	4.01	8.65	7.55	6.24
Pastures	34.45	37.95	38.22	13.43	2.89	3.04
Other	255.02	323.92	347.98	220.12	240.64	243.15
Total	952	952	952	357	357	357

Table 7. Vegetation cover in the municipalities of Kursumlija and Topola obtained from NDVI for 2006, 2011 and 2014.

Municipality	Municipality total area (km²)	Calculated on the basis of CLC for 2006	Calculated on the basis of CLC for 2012	Calculated on the basis of NDVI for 2006	Calculated on the basis of NDVI for 2011
Topola	357	57.99	57.69	52.14	51.24
Kursumlija	952	630.45	629.23	600.97	530.64

Table 8. Forest areas calculated on the basis of NDVI and CLC for the years 2006 and 2012 (in km²).

	CLC-NDVI difference (km²)		CLC-NDVI/NDVI difference (in %)	
Year	2006	2012	2006	2012
Topola	5.85	6.45	11.22	12.59
Kursumlija	29.48	98.59	4.90	18.58

Table 9. Difference between forest areas calculated on the basis of NDVI and CLC for the years 2006 and 2012 (in km² and in %).

	Calculated on the basis of CLC				Calculated on the basis of NDVI			
Municipality	Total forest area (km²)	Coniferous forest	Broad-leaved forest	Mixed forest	Total forest area (km²)	Coniferous forest	Broad-leaved forest	Mixed forest
Topola	57.69	0.15	55.54	2.00	51.24	0.17	49.05	2.02
Kursumlija	629.2	3.44	619.78	6.01	530.64	3.65	522.04	4.95

Table 10. Forest areas calculated on the basis of NDVI and CLC for 2012(2011) (in km²).

No.	Forest area obtained by NDVI (ha)	Forest area obtained by GPS (ha)	Deviation NDVI/GPS (%)
1	0.473	0.474	99.79
2	0.345	0.345	100.00
3	0.715	0.714	100.14
4	0.763	0.762	100.13
5	0.699	0.699	100.00
6	0.545	0.545	100.00
7	0.110	0.110	100.00
8	0.112	0.112	100.00
9	0.175	0.175	100.00
10	0.049	0.050	98.00
11	0.061	0.061	100.00
12	0.058	0.059	98.30
13	0.061	0.062	98.39
14	0.071	0.072	98.61
15	0.044	0.044	100.00
16	0.041	0.042	97.62
17	0.055	0.056	98.21
18	0.055	0.055	100.00
19	0.051	0.051	100.00
20	0.032	0.032	100.00
21	0.035	0.035	100.00
22	0.031	0.030	103.33
23	0.024	0.025	96.00
24	0.021	0.021	100.00
25	0.070	0.069	101.45
26	0.451	0.452	99.78
27	0.035	0.034	102.94
28	0.033	0.032	103.125
29	0.020	0.021	95.24
30	0.022	0.022	100.00
31	3.944	3.846	102.55
32	3.625	3.623	100.05

No.	Forest area obtained by NDVI (ha)	Forest area obtained by GPS (ha)	Deviation NDVI/GPS (%)
33	3.234	3.232	100.06
34	3.985	3.987	99.95
35	4.003	4.001	100.05
36	3.256	3.258	99.94
37	3.685	3.683	100.05
38	3.942	3.944	99.95
39	2.545	2.547	99.92
40	3.025	3.026	99.97
41	1.589	1.587	100.13
42	2.452	2.460	99.67
43	2.551	2.551	100.00
44	2.703	2.704	99.96
45	2.555	2.556	99.96
46	0.633	0.632	100.16
47	0.652	0.652	100.00
48	0.580	0.581	99.83
49	0.554	0.555	99.82
50	0.281	0.281	100.00
51	0.332	0.332	100.00
52	0.354	0.354	100.00
53	0.420	0.422	99.95
54	0.450	0.449	100.22
55	0.363	0.365	99.45
56	0.159	0.157	101.27
57	0.455	0.456	99.78
58	0.030	0.031	96.77
59	0.029	0.029	100.00
60	0.021	0.022	95.45
61	0.025	0.026	96.15
62	0.058	0.057	101.75
63	0.049	0.048	102.08
64	0.035	0.034	102.94

Table 11. Deviation between 64 forest areas/units obtained by NDVI and by GPS (for spatial units smaller than 4 ha) in municipality of Topola for 2014 (in %).

4. Discussion

Since, according to the official statistics, Kursumlija experienced a 10% (60 km^2) loss of the forest area in the 2006–2011 period, or, according to our NDVI results, 15.5% (93 km^2) loss in the 2006–2014 period, this is very clear case of alarmingly rapid process of *deforestation*, in very sharp contrast with very modest rates of deforestation in Serbia and in municipality of Topola. Numerous studies convincingly showed that this extremely quick process of *deforestation* in Kursumlija is the obvious result of illegal logging [3, 13–15, 37].

Although CLC have recently been used in Serbia for spatial planning at the local level, the main problem with CLC data is that (a) although CLC data are produced at various levels (international, national, regional, and local; [31, 32]), CLC is actually a predominant regional database, updated rarely (every 5–10 years), whereas NDVI is available every year and (b) NDVI is much more precise than CLC.

When official statistics were compared with NDVI and CLC forest areas for the same year (2006, 2011, 2012, and 2014), NDVI was more precise than CLC. Actually, the main problem with applying CLC for Serbia is that any (for CLC) unrecognizable spatial area smaller than 4 ha is automatically added to the closest recognizable larger area. This proved to be decisive for Serbia, where privately owned forest parcels, which account for half of the total forest area of the country, usually cover much smaller areas (the average private holding is 0.3 ha; [3]) and it is the main cause for CLCs (up to 15.6% imprecise) larger than official forest area estimates. In short, commonly used EU CORINE land cover (CLC), which serves as a tool for fulfilling pan-European monitoring needs [49] proved not to be very suitable for local forest management in Serbia (questionable results regarding forests were also determined in Slovenia, for example, [54]). In addition, apart from the obvious CLC imprecision for studies at the local level, CLC data are not available for every year.

On the other hand, all our NDVI results for Topola and for Kursumlija for 2006 and 2011 completely fit within the ±5% margin of error, while for 2014 they are much closer (±3% difference) to the "ground truth" (checked by our multiple GPS field surveys) and better fit the forest area trajectory. These discrepancies between our NDVI results and official inventory updates are obviously permanently growing: for Kursumlija, 2006 official forest inventory shows only 3.44 km^2 more forest areas; for 2011, 13.65 km^2 more; and for 2014, 38.62 km^2 more forest areas. Since our NDVI results for forest areas proved to be very close to the "ground truth" for 2014, it clearly indicates that in case of Serbia, official forest area updates are not precise. As Foody [56] points out, although remote sensing has been used successfully in mapping a range of land covers, land cover maps are often judged to be of insufficient quality for operational applications, especially when they are evaluated against some ground or other reference data set (like official forest area updates, in our case). Disagreements between the two data sets are typically interpreted as errors in the land cover map derived from the remotely sensed data [56].

Since our NDVI results for forest areas in 2014 proved to be very close to the "ground truth", these growing discrepancies between official forest area updates and our NDVI results are obviously the result of insufficient quality of official forest area updates.

Actually, there are several reasons that official forest area updates are very imprecise:

First, private forests are inventoried via assessment methods (cadastral data are used for assessing forest area) and Serbia is well known for the lack of cadaster updating.

Second, while national forest inventory in forests includes only inventory unit areas larger than 0.5 ha, private forests (that constitute 52% of the nation's forest resources) are characterized by very small plots (average size: 0.3 ha).

Third, official inventory of illegal loggings in private forests in Serbia is notoriously imprecise. The real amount of logged wood in private forests is six time as high as the registered amount [73].

Fourth, Serbian forest authorities have limited access to this extremely sensitive Serbia-Kosovo border area (formally under Serbian forest estates Vranje, Kursumlija, Leskovac, Raska, and Leposavic).

Also, it is important to underline here that none of the number of shortcomings that are usually addressed to the NDVI use [74, 75], proved to be relevant in the case of Kursumlija. Firstly, it is very easy to locate and quantify overall amounts of timber harvested in the case of Kursumlija municipality, because illegal logging produce large canopy gaps that go/extend from the border of Kosovo to approximately 3–4 km into the Kursumlija's territory. Secondly, from the field survey (hammer marks and size/height of stumps in the field), it is obvious that it is the clear case of illegal cutting, and thirdly, since illegal logging in Kursumlija is organized by groups of individuals, with market-oriented behavior [76] which are part of organized crime and closely tied to other criminal activities such as corruption, violence, and money laundering [77], it is, of course, not any sort of concession allocation issue.

Hence, it is exactly the objectivity of remote sensing that can be of the greatest help in resolving extremely quick and forceful process of illegal logging in this very sensitive southern Serbian area [12, 13, 76, 77], by providing an reliable, up-to-date alternative data source to quantify forest cover and change (independent of often very imprecise official governmental data sources) [78].

5. Conclusions

Through this analysis of NDVI and CLC results, CLC proved not to be a very suitable tool for local forest management in Serbia: (a) CLC is an example of a top-down process of European standardization that does not properly take into account unique individual features of post-socialist countries like Serbia, which resulted in extremely high (up to 15.6%) level of results imprecision and (b) CLC data are not available for every year. On the other hand, it is evident that NDVI, especially in southern Serbian municipalities with prevalent illegal logging can provide local forest managers with much precise annual information about forest area change.

Despite certain shortcomings [18, 24, 74, 75], classification and area estimation of various land cover types based on NDVI, has obviously advanced to a point where it surpasses old wood inventory techniques, especially in the case of post-socialist countries like Serbia. Specifically,

- It is relatively cheap [17] and quick, and it can provide forest managers with precise, up-to-date, annual information.
- It is easy to implement, which is of crucial importance for Serbia, where national forest inventories have been carried out very rarely. The last three national forest inventories were carried out at roughly 20-year intervals (in 1961, 1979, and 2007).

The objectivity of the method can significantly help in avoiding abrupt, unrecorded illegal logging that in post-socialist countries can exceed the legal harvest by a factor of 10.

Acknowledgements

This work was supported by the Ministry of Science and Technological Development of the Republic of Serbia under grant no. 37010.

Author details

Jovanović M. Miomir*, Miško M. Milanović and Bojan R. Vračarević

*Address all correspondence to: miomir.m.jovanovic@gmail.com

Faculty of Geography, University of Belgrade, Belgrade, Serbia

References

[1] World Bank. Forests Sourcebook: Practical guidance for sustaining forests in development cooperation. Washington: World Bank; 2008

[2] Ministry of agriculture forestry and water management of the Republic of Serbia. The National Forest Inventory of Serbia. Belgrade: Planeta Print; 2009

[3] Glavonjic B, Jovic D, Vasiljevic A, Kankaras R. Forest and forest products country profile: Serbia and Montenegro. Geneva: United Nations; 2005

[4] FAO. Global forest resources assessment update 2005. Terms and definitions (final version) forest resources assessment programme working paper 83/E. Rome, 2004

[5] Rominjin E, Ainembabazi JH, Wijaya A, Herold M, Angelsen A, Verchot L, et al. Exploring different forest definitions and their impact on developing REDD+ reference emission levels: A case study for Indonesia. Environmental Science & Policy. 2013;**33**:246-259

[6] GFOI. Integrating remote-sensing and ground-based observations for estimation of emissions and removals of greenhouse gases in forests. Geneva: Group on Earth Observations (GEO), 2013

[7] Kohl M, Traub B, Paivinen R. Harmonization and standardization in multi-national environmental statistics—mission impossible? Environmental Monitoring and Assessment. 2000;**63**(2):361-380

[8] Lawrence M, McRoberts RE, Tomppo E, Gschwantner T, Gabler K. Comparisons of national forest inventories. In: Tomppo E, Gschwantner T, REM ML, editors. National Forest Inventories. New York: Springer; 2010

[9] Nonić D, Petrović N, Medarević M, Glavonjić P, Nedeljković J, Stevanov M, Orlović S, Lj R, Djordjević I, NR PZ. Forest Land Ownership Change in Serbia. COST Action FP1201 FACESMAP Country Report. Vienna: European Forest Institute Central-East and South-East European Regional Office; 2015

[10] Taff GN, Müller D, Kuemmerle T, Ozdeneral E, Walsh SJ. Reforestation in Central and Eastern Europe after the breakdown of socialism. In: Nagendra H, Southworth J, editors. Reforesting Landscapes. New York: Springer; 2010. p. 121-147

[11] World Bank. Ensuring Sustainability of forests and livelihoods through improved governance and control of illegal logging for economies in transition. Washington D. C.: World Bank. 2005

[12] Harou P, Hajredini E. Illegal logging in Kosovo. Washington D. C.: USAID; 2009

[13] Jovanovic M, Milanovic M. Remote sensing and Forest conservation: Challenges of illegal logging in Kursumlija municipality (Serbia). In: Shukla G, Chakravarty S, editors. Forest Ecology and Conservation. Rijeka: InTech; 2017

[14] The Regional Environmental Center for Central and Eastern Europe. Illegal logging activities in the Republic of Serbia—B. Diagnostic Audit. Belgrade; 2009

[15] Regional Environmental Center. Illegal logging activities in the Republic of Serbia—A. Fact-finding study. Belgrade; 2009

[16] Blackman A. Evaluating forest conservation policies in developing countries using remote sensing data: An introduction and practical guide. Forest Policy and Economics. 2013;**34**:1-16

[17] Woodcock CE, Allen R, Anderson M, Belward A, Bindschadler R, Cohen WB, et al. Free access to Landsat data (Letter). Science. 2008;**80**:320

[18] Franklin SE. Remote sensing for sustainable forest management [Internet]. New York. 2001. 407 p. Available from: http://www.amazon.ca/exec/obidos/redirect?tag=citeulike 09-20&path=ASIN/1566703948

[19] Romijn E, Lantican CB, Herold M, Lindquist E, Ochieng R, Wijaya A, et al. Assessing change in national forest monitoring capacities of 99 tropical countries. Forest Ecology and Management. 2015;**352**:109-123

[20] MacDicken K. Global Forest resources assessment 2015: What, why and how? Forest Ecology and Management. 2015;**352**:3-8

[21] Jensen JR. Remote Sensing of the Environment: An Earth Resource Perspective. Upper Saddle River, NJ: Prentice Hall; 2007

[22] Weier J, Herring D. Measuring Vegetation (NDVI & EVI) [Internet]. Feature Articles. 2000. pp. 1-4. Available from: http://earthobservatory.nasa.gov/Features/Measuring Vegetation/measuring_vegetation_1.php

[23] Chen CH. In: Image Processing for Remote Sensing. New York: CRC Press; 2008. 400 p

[24] Campbell J, Wynne R. Introduction to Remote Sensing. New York: Guilford Press; 2011

[25] Win RN, Suzuki R, Takeda S. Remote sensing analysis of Forest damage by selection logging in the Kabaung reserved forest, Bago Mountains Myanmar. Journal of Forest Research. 2012;**17**(2):121-128

[26] Montandon LM, Small EE. The impact of soil reflectance on the quantification of the green vegetation fraction from NDVI. Remote Sensing of Environment. 2008;**112**(4):1835-1845

[27] Vohland M, Stoffels J, Hau C, Schüler G. Remote sensing techniques for forest parameter assessment: Multispectral classification and linear spectral mixture analysis. Silva Fennica. 2007;**41**(3):441-456

[28] De Jong, S. M., Van der Meer FD. Remote sensing image analysis: Including the spatial domain. New York: Springer; 2005

[29] Wang Q, Tenhunen JD. Vegetation mapping with multitemporal NDVI in north eastern China transect (NECT). International Journal of Applied Earth Observation and Geoinformation. 2004;**6**(1):17-31

[30] Lillesand TM, Kiefer RW, Chipman JW. Remote sensing and image interpretation [Internet]. Vol. 3, New York, Chichester, Brisbane, Toronto: 6IS s, 2004, 756 p. Available from: http://www.osti.gov/energycitations/product.biblio.jsp?osti_id=6028047

[31] Bossard M, Feranec J, Otahel J. CORINE Land Cover technical guide—addendum 2000. Technical report 40 [Internet]. 2000 [cited 2016 May 17]. Available from: http://www.eea.europa.eu/publications/tech40add

[32] Neumann K, Herold M, Hartley A, Schmullius C. Comparative assessment of CORINE2000 and GLC2000: Spatial analysis of land cover data for Europe. International Journal of Applied Earth Observation and Geoinformation. 2007;**9**(4):425-437

[33] Heymann Y, Steenmans CH, Croissille G, Bossard M. CORINE land cover, Technical guide. Luxembourg: office for official Publications of the European Communities; 1994

[34] Perdigäo V, Annoni A. Technical and methodological guide for updating CORINE land cover data base [Internet]. 1997 [cited 2014 Jan 1]. Available from: http://www.ec-gis.org/docs/F27057/CORINE.PDF

[35] Genovese G, Vignolles C, Negre T, Passera GA. Methodology for a combined use of normalised difference vegetation index and CORINE land cover data for crop yield monitoring and forecasting: A case study on Spain. Agronomie. 2001;**21**(1)

[36] Potapov PV, Turubanova SA, Tyukavina A, Krylov AM, McCarty JL, Radeloff VC, et al. Eastern Europe's forest cover dynamics from 1985 to 2012 quantified from the full Landsat archive. Remote Sensing of Environment. 2015;**159**:28-43

[37] Jovanovic M, Milanovic M, Zorn M. The use of NDVI and CORINE land cover databases for forest management in Serbia. Acta Geographica Slovenica. 2018;**58**(1):109-123

[38] Jovanović M, Milanović M. Normalized difference vegetation index (NDVI) as the basis for local Forest management. Example of the municipality of Topola. Polish Journal of Environmental Studies. 2015;**24**(2):529-535

[39] Verstraete MM, Pinty B. Designing optimal spectral indexes for remote sensing applications. IEEE Transactions on Geoscience and Remote Sensing. 1996;**34**(5):1254-1265

[40] Chávez PSJ. Image-based atmospheric corrections—revisited and improved. Photogrammetric Engineering & Remote Sensing. 1996;**62**(9):1025-1036

[41] Hájek F. Process-based approach to automated classification of forest structures using medium format digital aerial photos and ancillary GIS information. European Journal of Forest Research. 2008;**127**(2):115-124

[42] Johnson LF, Trout TJ. Satellite NDVI assisted monitoring of vegetable crop evapotranspiration in california's san Joaquin Valley. Remote Sensing. 2012;**4**(2):439-455

[43] Xiao X, Boles S, Liu J, Zhuang D, Liu M. Characterization of forest types in northeastern China, using multi-temporal SPOT-4 VEGETATION sensor data. Remote Sensing of Environment. 2002;**82**:335-348

[44] Myneni RB, Tucker CJ, Asrar G, Keeling CD. Interannual variations in satellite-sensed vegetation index data from 1981 to 1991. Journal of Geophysical Research. 1998;**103**:6145-6160

[45] Jenkins JP, Braswell BH, Frolking SE, Aber JA. Predicting spatial and interannual patterns of temperate forest springtime phenology in the eastern U.S. Geophysical Research Letters. 2002;**29**(24)

[46] Ofosu Anim D, Tiereyangn Kabo-bah A, Nti Nkrumah P, Tackmore Murava R. Evaluation of NDVI using SPOT-5 satellite data for Northern Ghana. Environmental Management and Sustainable Development. 2013;**2**(1)

[47] Roni R. Surface temperature and NDVI generation and relation between them: Application of remote sensing. Asian Journal of Engineering and Technology Innovation. 2013; **1**(1):8-13

[48] Di Gregorio A, O'Brien D. Overview of land-cover classifications and their interoperability. In: Giri C, editor. Remote Sensing of Land Use and Land Cover. London, NewYork: CRC Press; 2012

[49] Jansen LJM, Badea A, Milenov P, Moise C, Vassilev V, Milenova L, et al. The use of the land-cover classification system in eastern European countries: Experiences, lessons learnt and the way forward. In: Manakos I, Braun M, editors. Land Use and Land Cover Mapping in Europe. London, NewYork: Springer; 2014

[50] Hord RM. Digital image processing of remotely sensed data [Internet]. Enschede; 1982. Available from: https://books.google.co.in/books?id=e-tRAAAAMAAJ

[51] Büttner G, Kleeschulte S. Implementation of CLC2000 in Serbia-Montenegro—technical annex. 2005

[52] EEA. CLC2006 Technical Guidelines [Internet]. 2007 [cited 2014 Jan 1]. Available from: http://www.eea.europa.eu/publications/technical_report_2007_17

[53] Hočevar M, Kobler A. Uporaba podatkov CORINE land cover za izgradnjo informacijskega sistema o gozdovih na državni ravni. Gozdarski Vestnik. 2001;**46**(3):353-369

[54] Gabrovec M, Petek F. V Sloveniji vedno manj gozda? (Ne)uporabnost metodologije CORINE Land Cover. Delo. 2004;**46**(303)

[55] Stehman SV, Czaplewski RL. Design and analysis for thematic map accuracy assessment. Fundamental principles. Remote Sensing of Environment. 1998;**64**:331-344

[56] Foody GM. Status of land cover classification accuracy assessment. Remote Sensing of Environment. 2002;**80**:185-201

[57] Czaplewski RL. Accuracy assessment of maps of forest condition: Statistical design and methodological considerations. In: Wulder MA, Franklin SE, editors. Remote Sensing of Forest Environments: Concepts and Case Studies. Boston: Kluwer; 2003. p. 115-140

[58] Strahler AH, Boschetti L, Foody GM, Friedl MA, Hansen MC, Herold M, et al. Global Land Cover Validation: Recommendations for Evaluation and Accuracy Assessment of Global Land Cover Maps. Ispra; 2006

[59] Stehman SV, Foody GM. Accuracy assessment. In: Warner TA, Nellis MD, Foody GM, editors. The SAGE Handbook of Remote Sensing. London: Sage; 2009. p. 297-309

[60] McRoberts R, Cohen WB, Næsset E, Stehman SV, Tomppo EO. Using remotely sensed data to construct and assess forest attribute maps and related spatial products. Scandinavian Journal of Forest Research. 2010;**25**:340-367

[61] Titshall LW, O'Connor TG, Morris CD. Effect of long-term exclusion of fire and herbivory on the soils and vegetation of sour grassland. African Journal of Range & Forage Science. 2000;**17**:70-80

[62] Wood SW, Murphy BP, Bowman DM. Firescape ecology: How topography determines the contrasting distribution of fire and rain forest in the south-west of the tasmanian wilderness world heritage area. Journal of Biogeography. 2011;**38**:1807-1820

[63] Hwang T, Song C, Vose JM, Band LE. Topography-mediated controls on local vegetation phenology estimated from modis vegetation index. Landscape Ecology. 2011:541-556

[64] Kayiranga A, Ndayisaba F, Nahayo L, Karamage F, Nsengiyumva JB, Mupenzi C, et al. Analysis of climate and topography impacts on the spatial distribution of vegetation in the Virunga volcanoes massif of east-Central Africa. Geosciences. 2017;**7**:17

[65] Day FP, Monk CD. Vegetation patterns on a southern Appalachian watershed. Ecology. 1974;**55**:1064-1074

[66] Busing RT, White PS, MacKende MD. Gradient analysis of old spruce-fir forest of the great Smokey Mountains circa 1935. Canadian Journal of Botany. 1992;**71**:951-958

[67] Jin XM, Zhang Y-K, Schaepman ME, Clevers JGPW, Impact SZ. Of elevation and aspect on the spatial distribution of vegetation in the qilian mountain area with remote sensing data. The International Archives of the Photogrammetry, Remote Sensing and Spatial Information Sciences. 2008;**37**:1385-1390

[68] Statistical Office of the Republic of Serbia. Municipalities and Regions of the Republic of Serbia 2012. Belgrade: Statistical Office of the Republic of Serbia; 2013

[69] Statistical Office of the Republic of Serbia. Municipalities and Regions of the Republic of Serbia 2007. Belgrade: Statistical Office of the Republic of Serbia; 2008

[70] Statistical Office of the Republic of Serbia. Municipalities and regions of the republic of Serbia 2015. Belgrade: Statistical Office of the Republic of Serbia; 2016

[71] Lunetta RS, Knight JF, Ediriwickrema J, Lyon JG, Worthy LD. Land-cover change detection using multi-temporal MODIS NDVI data. Remote Sensing of Environment. 2006;**105**(2):142-154

[72] Eastman R. IDRISI32 Release 2—Tutorial. Worcester: Clark Labs; 2001

[73] Markus-Johansson M, Mesquita B, Nemeth A, Dimovski M, Monnier C, Kiss-Parciu P. Illegal Logging in South Eastern Europe. Szentendre: The Regional Environmental Center for Central and Eastern Europe. 2010

[74] Khai TC, Mizoue N, Kajisa T, Ota T, Yoshida S. Stand structure, composition and illegal logging in selectively logged production forests of Myanmar: Comparison of two compartments subject to different cutting frequency. Global Ecology and Conservation. 2016;**7**:132-140. Available from: http://dx.doi.org/10.1016/j.gecco.2016.06.001

[75] Lawson S, MacFaul L. Illegal logging and related trade [Internet]. Chatham House, 2010, pp. 1-154. Available from: papers2://publication/uuid/C7E061BB-61C3-4A58-A30D-DF4FFDA3CBD2

[76] Bouriaud L, Nichiforel L, Nunes L, Pereira H, Bajraktari A. A property rights-based analysis of the illegal logging for fuelwood in Kosovo. Biomass and Bioenergy. 2014; **67**:425-434

[77] WWF. Illegal logging & The EU. An analysis of the EU export & import market of illegal wood and related products. Brussel: WWF European Policy Office; 2008

[78] Sundstrom A. Understanding illegality and corruption in forest governance. Journal of Environmental Management. 2016

Chapter 2

Changes in Land Management of Iberian Rangelands and Grasslands in the Last 60 Years and their Effect on Vegetation

Manuel Pulido Fernández,
Joaquín Francisco Lavado Contador,
Susanne Schnabel, Álvaro Gómez Gutiérrez and
Javier Lozano-Parra

Additional information is available at the end of the chapter

http://dx.doi.org/10.5772/intechopen.72490

Abstract

Rangelands and grasslands occupy now more than 3 million ha in the Iberian Peninsula representing one of the most valuable ecosystems. They are particularly interesting due to their great geographical spread and heterogeneity in land management. Significant changes have been undergone in the last 60 years affecting vegetation. The main goal in this study was to compile existing information about the changes in the Iberian rangelands and grasslands, their geographical distribution, typologies, main characteristics as well as past and recent land management based on a thorough bibliographical review and serious research. This has been subdivided into five large groups according to climate and human drivers: (1) Mediterranean rangelands and steppes, (2) oceanic grasslands, (3) mountainous meadows, (4) semi-arid steppes and (5) Portuguese rangelands and grasslands. Two milestones over the past 60 years were remarkable as main driving forces: rural exodus in the 1960s and 1970s and Spain and Portugal joining the European Union in 1986. They have provoked both processes of intensification and extensification at the same time on different scales. Many farms have been progressively fragmented using wire fences, and the numbers of livestock have been significantly increased. Land abandonment and grazing exclusion have provoked a large shrub encroachment of species such as *Retama sphaerocarpa* or *Cistus ladanifer*.

Keywords: EU subsidies, fencing, grazing, land abandonment, shrub encroachment

1. Introduction

A long period of time has passed since the Latin geographer Strabo in his book *Geographiké* described how Roman Hispania could be crossed by a squirrel jumping from tree to tree from north to south [1]. Many centuries of intense human activity have led to the progressive reduction of this original dense forest converted into agricultural land, grazing areas or settlement of the current cities [2]. Of particular interest are those lands that have been progressively cleared aiming to obtain pastures, either by removing all the trees (grasslands) or by combining scattered trees with the management of a productive herbaceous layer (rangelands), to feed domestic animals [3].

These rangelands, called *dehesas* in Spanish and *montados* in Portuguese, occupy more than 2 million ha in the southwestern part of the Iberian Peninsula (this value can vary depending on the author consulted). They represent one of the most valuable ecosystems for both countries (Spain and Portugal) for different reasons: biodiversity, meat production, landscape, preservation of traditional values and rural population fixation [4]. Grasslands are also very important in terms of the area they occupy (more than 1 million ha in size), but above all they are particularly interesting due to their great geographical spread and heterogeneity in land management [5].

The origin of these grasslands is probably the same for the better-known rangelands (dehesas and montados): the clearing of the former Mediterranean forest in order to obtain pasture for livestock, dominated by Quercus genus species: holm and cork oak. López Sáez et al. [6] attribute a prehistoric origin to this type of practice, but it is not until the reconquest of the territory under Arab domination by the Christian kingdoms (Portugal, Leon, Castile and Aragon) in the Middle Ages (1212–1492) that the current landscape of the Iberian rangelands and grasslands begin to take shape [7]. However, Llorente Pinto [8] dates the origin of these land systems to the eighteenth century.

Although the northern part of the Iberian Peninsula, dominated by an oceanic climate type [9], and those regions located close to the Mediterranean Sea have experienced the same process of clearing of the original forest, land management has been historically different in these areas [10], and in Portugal as well [11], due to many factors such as climate, topography or land tenure. Nevertheless, the most significant changes in land management have been undergone over the past 60 years when some areas of Portugal and Spain began an economic process of industrialization [12] and the rules of the European Union (EU) Common Agricultural Policy (CAP) came into effect [13].

This study aims at compiling further much-needed information about Iberian rangelands and grasslands: geographical distribution, typologies and their main characteristics. Finally, past and recent land management has been analysed based on a thorough bibliographical review and serious research in order to better understand how these changes have affected vegetation (e.g., composition, structure or species) on a landscape scale.

2. Geographical distribution

Figure 1 shows the geographical distribution of grasslands throughout the Iberian Peninsula. They are distributed in soils developed on different rock types: siliceous (slate and granite),

http://dx.doi.org/10.5772/intechopen.72490

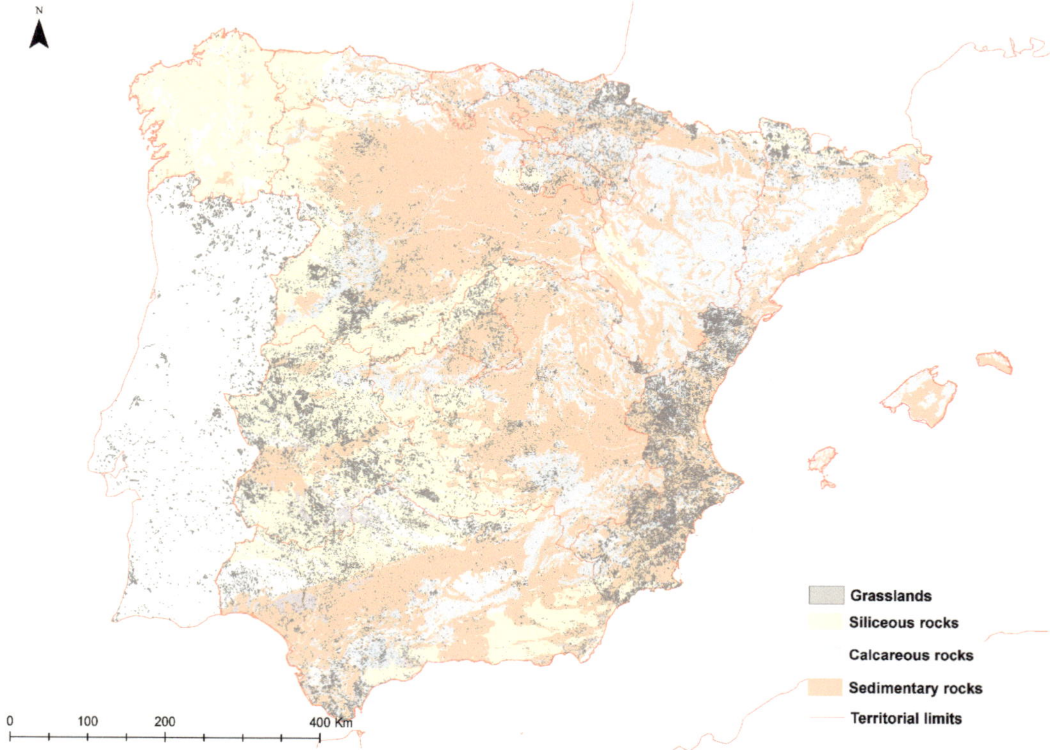

Figure 1. Distribution of grasslands in the Iberian peninsula. Data source: IGME [14], IGN [15] and CLC [16].

calcareous (limestone and dolomite) and tertiary sediments [14]. From a political point of view, the presence of grasslands is a common feature in most of the 15 autonomous regions into which peninsular Spain is divided, being particularly important in regions such as Valencia and Murcia (East) and Extremadura (West) [15]. In Portugal, around 175,000 ha of grasslands are disseminated throughout the country being mostly concentrated in the northern and western parts along the Spanish border [16].

3. Types of rangelands and grasslands

Iberian rangelands and grasslands are mostly represented by annual species, in areas thought to have been historically dominated by perennial forms, mainly as a consequence of the seasonality of rainfall [17]. The diversity of climate types (updated classification by Köppen-Geiger) is the first criterion that explains the existing differences between rangelands and grasslands in this region. Broadly speaking, the southern half of the Iberian Peninsula is dominated by a Mediterranean climate (Csa) with dry and warm summers, the northern half by a Mediterranean oceanic climate (Csb) with milder summers, mountainous areas with oceanic climate (Cfb) and SE Spain with a semi-arid climate (Bsh) [18] (**Figure 2**).

The description of Spanish rangelands and grasslands and their management is therefore divided into 4 groups: (3.1.) Mediterranean rangelands and steppes, (3.2.) oceanic grasslands,

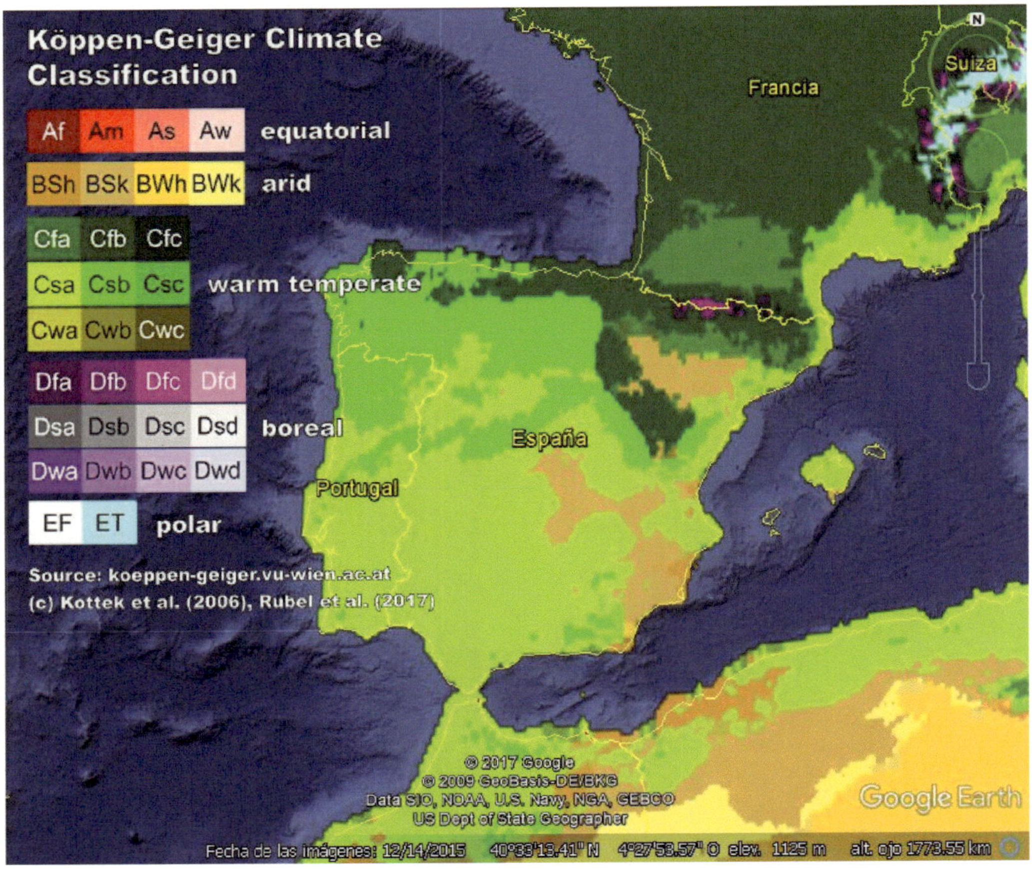

Figure 2. Köppen-Geiger's updated climate classification of the Iberian peninsula. Source: [18].

(3.3.) mountainous meadows and (3.4.) semi-arid steppes. (3.5.) Portuguese rangelands and grasslands are treated as a separated group. Relevant information of their land management (past and present) as well as of their vegetation features (including illustrative pictures) of each case is provided in the next sub-chapters.

3.1. Mediterranean rangelands and steppes

Mediterranean grasslands in Spain are a consequence of a more advanced state of tree degradation than in rangelands (dehesas). They have arisen due to a total deforestation in order to obtain croplands for cereal cultivation at some point in history. In Spain, tree clearing has been constant since the Middle Ages, but there is evidence that both the nineteenth and twentieth centuries were the periods of highest intensity. The confiscation of land owned by the Catholic Church (1798–1856), economic consequences of the Spanish Civil War (1936–1939) and a doubling in population (1860–1890) were the main driving forces [19].

Dehesas (**Figure 3** left) occupy more than two million ha in SW Spain [20] providing many ecosystem services [21]. Their landscape is characterised by the presence of scattered trees

Figure 3. Illustrative pictures of a dehesa (left) and a pastizal farm (right). Author: Javier Lozano-Parra.

belonging to the Quercus genus (holm and cork oak) used to obtain direct benefits such as fruit (acorn for feeding animals), wood or cork, and indirect benefits like protection against wind and sun (shadow in summer), soil erosion or nutrient pumping [22]. These trees enhance the landscape's heterogeneity increasing its complexity and biodiversity [23]. When most of the trees are removed (cultivation or lack of regeneration), these lands are converted into a treeless rangeland or grassland, commonly known in Spanish as *pastizales*, that could be actually considered as a steppe (**Figure 3** right).

The steppe is the most common landscape in Mediterranean and semi-arid grasslands. They are currently managed as grazing lands where annual species of legumes (*Ornithopus compressus L.* and several species of Trifolium), grasses (*Aira caryophyllea* L., *Bromus sp.*, etc.) and other families are dominant [24] (**Figure 4** left). In many cases, these grasslands are still rotationally cultivated every 3 or 4 years (traditional practice) with fodder species such as oats (*Avena sativa*) and vetch (*Vicia sativa*) [25]. Contrariwise, many of them are also abandoned giving rise to shrub encroachment (matorralization) of species such as *Cistus ladanifer*, *Lavandula stoechas* and *Retama sphaerocarpa* [26] (**Figure 4** right).

Figure 4. Examples of a traditional grassland farm where rotational cultivation is still practiced (left) and another farm abandoned over the past 20 years. Author: Javier Lozano-Parra.

3.2. Oceanic grasslands

Grasslands both in northern Spain and in high mountainous areas are less abundant than in the southern half of the country (Mediterranean and semi-arid). It is quite usual to consider mountain meadows with evergreen grasses, grazed mainly by cattle, as the typical grassland of this geographical region, particularly in the Pyrenees [27] and in the Cantabrian Mountains [28]. In addition, some regions such as Galicia (NW Spain) are relatively densely populated (92 people km^{-2}), disperse settlements (a lower human habitat concentration) and micro-properties being the dominant features that have produced the complex mosaic that conforms this typical landscape [29] (**Figure 5**). Grasslands in the strictest sense (not to be confused with meadows) are mainly located in the Navarre region (western French border) although the dominance of Mediterranean climate in some areas (south) reduces its geographical distribution.

The Spanish region of Navarre (ca. 10,400 km^2) has an important rainfall gradient (350–2500 mm $yr.^{-1}$) that allows for the coexistence of dry [30] (**Figure 6**) and wet grasslands [31] (**Figure 7**) as well as mountain meadows (**Figure 8**), rangelands and rich agricultural fields. Berastegi et al. [32] classified its grasslands (and meadows) in 7 major groups, 37 subgroups and 69 different types (including associations). They gathered around 60% of the number of species found in the region and 27 out of 37 subgroups belong to the list of Habitats of Common Interest of the European Union. The main land-use is sheep grazing since Roman times [33] in the driest grasslands [34] and cattle as elevation rises [35].

3.3. Mountainous meadows

Iberian mountainous meadows can be included within the *nemoral forest* class (vegetation zone) by Mucina et al. [36] or as alpine grassland (phytosociological class) by Rivas Martínez et al. [37]. They are represented by acidophilous grasslands (*CT (TRI)—Juncetea trifidi Hadač in Klika et Hadač 1944*) in the Pyrenees and in the Cantabrian Mountains although the presence of calcicolous grasslands (*CU (SES)—Elyno-Seslerietea Br.-Bl. 1948*) is also noteworthy. Other

Figure 5. Illustrative picture of the typical fragmented landscape of Galicia. Place: Concello de Abegondo, A Coruña. Authors: Javier Pulido Fernández and Mónica Cortijo Blanco.

Figure 6. Shrub formations of *Lygeum spartum* and *Rosmarinus officinalis* in Bárdenas Reales region (southern Navarre). Author: Rosa María Canals Tresserras.

Figure 7. Atlantic meadows cultivated in a farm in Roncesvalles (northern Navarre). Fodder cultivation (ryegrass) is grazed by an autochthonous sheep breed named Latxa. Author: Rosa María Canals Tresserras.

Figure 8. Mountain meadow grazed by sheep (raso breed) located in the Salazar valley (northern Navarre). Author: Rosa María Canals Tresserras.

mountainous systems, such as *Sistema Central* (Madrid and Castilla y León), *Sierra Nevada* (Granada, Andalusia) and *Serra da Estrela* (Central Portugal), show differences only in regard to local species, but they could be included within the same phytosociological classes.

The Pyrenees are dominated by acidophilous species such as *Carex curvula subsp. curvula* and *Festuca airoides Lam.* and calcicolous species such as *Oxytropis halleri subsp. halleri* and *O. pyrenaica*. Their natural pastures are mainly grazed by local breeds of cattle (**Figure 9**) although a human process of land abandonment has been witnessed for many years [38]. In the Cantabrian Mountains, *Oreochloa disticha subsp. blanka* is one of the most representative species [39]. Other peninsular mountainous systems are remarkable for the presence of endemism, particularly of vascular flora [40]. Nevertheless, the abandonment of traditional activities linked to the mountains (grazing, agricultural, wood harvesting or forestry) due to emigration from rural areas to industrial regions is a common feature throughout Spain, particularly in the 1960s and 1970s [41].

3.4. Semi-arid steppes

In the semi-arid areas of SE Spain, steppes are a common feature as well. They are mainly characterised by the presence of species such as *Stipa tenacissima*, *Thymus vulgaris*, *Rosmarinus officinalis* or *Lavandula angustifolia*. Their grasses are usually little developed (low height), and it is very common to find areas of uncovered soil producing a desert-like landscape as a result [42] (**Figure 10**). It contrasts with Mediterranean steppes, where bare soil surfaces only occur under conditions of heavy grazing and dry grasses are regularly visible in the summertime, feeding domestic animals such as *merina* breed sheep [43]. This landscape is also a common feature in southern Portugal (Algarve region) where the climate is relatively drier than in the rest of Portugal [10].

These grasslands have historically been used to harvest esparto fibre up until the decade of the 1970s [10]. Their natural pastures have been mainly grazed by goats [44]. Overgrazing is addressed as a recurrent practice in the past [42], but nowadays this kind of grasslands are being abandoned provoking problems of matorralization and wildfires [45]. In dry regions such as Murcia, these grasslands have been rotationally cultivated with rainfed fodder cereal taking advantage of relatively deeper soil at the valley bottoms (**Figure 11**) although they are

Figure 9. Local cattle breed grazing in a mountain meadow located on the northern face of sierra Bernera (central Pyrenees). Author: Teodoro Lasanta Martínez.

http://dx.doi.org/10.5772/intechopen.72490

Figure 10. Semi-arid grassland dominated by *Stipa tenacissima* in the Tabernas desert (Almería). Author: Isabel Miralles Mellado.

Figure 11. Typical rainfed cultivation of cereal in a grassland farm located in SE Spain. Place: Archivel, NW Murcia. Author: María Nieves García Marín.

Figure 12. Former tree orchards and croplands abandoned that are currently grazed by sheep in eastern Spain (Valencia region). Author: Artemi Cerdà.

being converted into a more intensive land-use [46]. Contrariwise, in the Valencia region, many tree orchards (citrus) are being progressively abandoned and consequently converted into grazing areas [47] (**Figure 12**).

3.5. Rangelands and grasslands in Portugal

Portugal and Spain are divided into two climate zones (Mediterranean and oceanic). The southern half of the country is also dominated by grasslands that are currently treeless rangelands (montados) and have a steppe-like landscape [48].

The main land-use in these rangelands and grasslands is obviously grazing combined with rotational crops of cereals or fodder species in some cases [49]. Grasslands dominated by *Stipa tenacissima* are also present in southern Portugal (The Algarve region) [10]. These facts guarantee a visual continuity of land systems on both sides of the Portuguese-Spanish border in spite of being land belonging to different countries [50] (**Figure 13**).

The northern half of Portugal has quite distinct grasslands as compared with Spanish grasslands than in the southern half of the country where the landscape between both countries presents a larger spatial continuity. An exception could be the mosaic landscape of Galicia (NW Spain) showing similarities with those of northern Portugal even in the inner provinces (e.g., Lugo). **Figure 14** shows the typical landscape of many parts of Northern Portugal and the region of Galicia. It is formed by a natural forest with differing degrees of human intervention, areas reforested by pine trees in the twentieth century and natural pastures. Nevertheless, some of them deserve to be highlighted due to their current situation.

The endangered system called in Portuguese 'campo-bouça' (**Figure 15**) is formed by a mosaic-like tapestry of grasslands surrounded by fruit orchards and forest (e.g., cork oak). It is grazed by cattle and typical of the northern region of Portugal close to the Minho and Douro rivers. Another important grassland type is called 'lameiros' in Portuguese (**Figure 16**). They are

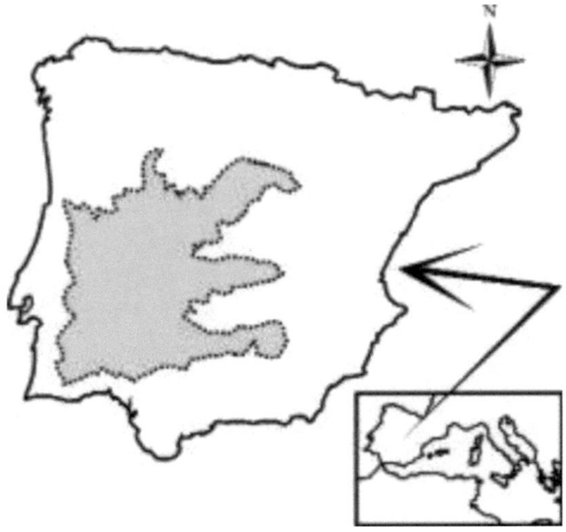

Figure 13. Broad distribution of dehesas and montados in the Iberian peninsula. Extracted from Gea-Izquierdo et al. [50], p. 342.

Figure 14. Typical landscape of inner Galicia. Place: Friol, Lugo. Author: Urbano Fra Paleo.

Figure 15. Image of a 'campo-bouça' located in Ponte de lima, Portugal. Author: Bruno Mateus.

Figure 16. Image of a 'lameiro' located in Rio de Onor, Bragança. Author: Carlos Aguiar.

traditional grasslands (even meadows) located in the high mountains in NE Portugal. Some autochthonous breeds of cows graze permanently on their evergreen grasses. These grasslands and their traditional management are a feature of the idiosyncrasy of local people, and they are at risk of disappearing [51].

4. Changes in land management

Spain and Portugal have undergone similar political and socioeconomic processes during the twentieth century. Both countries were governed by a dictatorial system: the dictatorships of General Franco in Spain (1936–1975) and Oliveira Salazar in Portugal (Estado Novo: 1933–1974). Their economic take-off happened in the 1960s and 1970s involving a strong emigration from little developed rural areas to the more developed industrial areas located within their own countries (Madrid, Catalonia, The Basque Country, Porto or Lisbon) or in other European countries (France, Germany, Switzerland, Luxembourg, The United Kingdom or The Netherlands) [52].

Figure 17 shows the evolution of the percentage of active workers for each economic sector in Spain during the twentieth century. Agricultural activities ceased to be dominant in the decade of the 1960s when a strong industrial development occurred along with a higher demand for services in the increasingly more populated cities [53]. This fact was responsible for the current depopulation of rural areas dominated by agriculture and the subsequent land abandonment or land-use intensification brought about by the introduction of agricultural machinery [54]. Parallel to this geographical process, national conversion plans for large-size rainfed croplands to irrigation [55] and national forestry policies for reforestation using eucalyptus and pine trees were prevalent in these rural areas [56].

Rangelands and grasslands have passed through different phases of land-use intensification induced by many diverse driving forces. After the Spanish Civil War (1936–1939), Spain went through a period of re-ruralisation, which resulted in the current conversion into croplands of thousands of hectares of land [57]. Trees and shrubs were increasingly cleared, and 4-year rotational cultivations of rye, wheat or barley were common throughout these years. Wood harvesting of holm and cork oak for heating (e.g., *picón*) and heavy grazing in many areas lead to processes of land degradation and soil erosion [58].

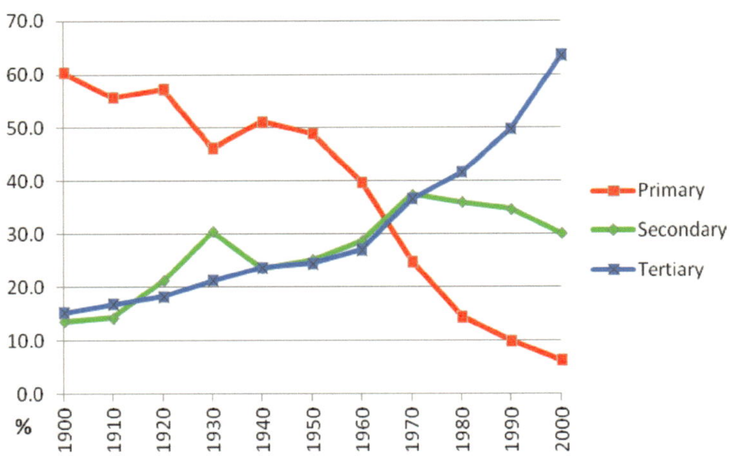

Figure 17. Evolution of the percentage of active workers by economic sector in Spain in the twentieth century. Data source: Spanish Statistical Institute (www.ine.es).

The decades of the 1960s and 1970s are key for understanding the period of decadence of many rangelands. Labour workers moved from rural settlements to industrial areas and those who remained in the rural areas began to demand higher salaries. The response of many owners was to buy some machinery (e.g., tractors or mowers) for agricultural labours and farms began to be fragmented using wire fences in order to reduce costs in shepherding and animal caring. In addition, African swine fever considerably reduced the income produced by the selling of Iberian pigs and their products [59].

This period of rangeland crisis was an influential factor in the advent of large-surface commercial fodder cultivations and in processes of land abandonment followed by shrub encroachment (matorralization). Many of these rangelands and grasslands were afforested by pine or eucalyptus trees following national programmes and others were converted into irrigation croplands for tomatoes, corn or cotton. Campos Palacín [60] counted a disappearance of more than 700,000 ha of rangelands between 1955 and 1981 converted into reforestations of croplands.

Spain and Portugal's joining the European Union meant their acceptance of the Common Agricultural Policy (CAP) rules and their subsidies promoting a higher number of animals. From 1986 to 2000, average animal stocking rates increased from 0.40 AU ha^{-1} to 0.70 AU ha^{-1} [61]. This increase in the number of animals was one of the main causes for the increase in the number of fenced areas in farms, subsequently reducing the connectivity between vegetation patches [62]. Another little-studied influential factor on vegetation was the progressive replacement of cattle for sheep [63].

Many grasslands and rangelands are being progressively occupied by shrub species such as *Retama sphaerocarpa* and *Cistus ladanifer* as a consequence of undergrazing and the lack of influence of goats within the grazing system (**Figure 18**). Nevertheless, Moreno and Rolo [64] concluded that this natural process of shrub encroachment is necessary to facilitate tree regeneration since the lack of trees is becoming a serious problem in rangelands of many parts of the Iberian Peninsula [65]. CAP agri-environmental subsidies are also promoting the seeding of different species or the exclusion of grazing in some areas in order to favour bird habitat protection [66].

Figure 18. Undergrazed fenced areas within a rangeland farm colonised by *Retama sphaerocarpa* over the past 20 years. Place: Monroy, Cáceres, Spain. Author: Javier Lozano-Parra.

Figure 19. Southern face of the sierra de las Corchuelas in Monfragüe National Park (Extremadura, Spain). The Tagus River splits a typical landscape of dehesa and a eucalyptus reforestation from the 1960s. Author: María José Mena Martín de Prado.

Finally, one of the most important changes in terms of vegetation has been the introduction of exotic tree species such as *Eucalyptus globulus*, *E. camaldulensis* or *Pinus halepensis* during the twentieth century, particularly in the northern part of the Iberian Peninsula and in Portugal as well (**Figure 19**). From 1941 to 1959, more than 1.3 million ha of land were reforested in order to reduce the wood imports following the instruction given by the National Plan for Reforestation of 1939 [67]. Madrigal [68] provided values of 3.7 million ha of reforested land up until 1986 in Spain, the 1970s showing particular importance after the reform of the former Spanish Ministry of Agriculture in 1971. Many of these eucalypti were planted for cellulose extraction [69].

5. Conclusions

The Iberian Peninsula is a territory that has been strongly intervened by human land-use since before Roman colonisation. This historical landscape modelling by human activities has shaped original land systems such as *dehesas* or *montados* (rangelands), steppes, micro-properties or endangered systems such as *campo-bouça* or *lameiros* in Portugal. Nevertheless, changes promoted by national governments (plans for reforestation or conversion into irrigated cropland), a significant increase in the secondary and tertiary economic sectors (rural exodus, land abandonment, mechanisation, etc.) in the 1960s or the application of the European Union Common Agricultural Policy, have motivated different types of vegetation. Traditional cereal fodder crops have been progressively replaced by natural pastures or commercial crops. Vegetation habitats have been increasingly fragmented, and processes of intensification and extensification have provoked changes on a farm scale (shrub encroachment, nitrophilous plants, etc.). Finally, this study is just a *bouquet* of the many land systems distributed throughout the Iberian Peninsula. Further, specific and empirical studies are still needed on this matter.

Acknowledgements

This work has been financed by the research projects CGL2014-54822R and IB16052 funded by the Spanish Ministry of Economy and Competitiveness and the Government of Extremadura, respectively. The payment of the chapter-processing charge has been made using economic funds from the CGL2014-54822R research project. Financial support was also offered by the Government of Extremadura and the European Regional Development Fund (Reference GR15032). We wish to thank Prof. João Azevedo and the authors of each illustrative picture.

Author details

Manuel Pulido Fernández[1*], Joaquín Francisco Lavado Contador[1], Susanne Schnabel[1], Álvaro Gómez Gutiérrez[1] and Javier Lozano-Parra[2]

*Address all correspondence to: mapulidof@unex.es

1 GeoEnvironmental Research Group, University of Extremadura, Cáceres, Spain

2 Department of Physical Geography, Pontifical Catholic University of Chile, Santiago, Chile

References

[1] Ollero HS, van Staalduinen MA. Iberian steppes. In: Werger MJA, van Staalduinen MA, editors. Eurasian Steppes. Ecological Problems and Livelihoods in a Changing World. Dordrecht: Springer Netherlands; 2012. pp. 273-286

[2] Barbero M, Bonin G, Loisel R, Quézel P. Changes and disturbances of forest ecosystems caused by human activities in the western part of the Mediterranean basin. Vegetatio. 1990;**87**:151-173

[3] Plieninger T, Pulido FJ, Konold W. Effects of land-use history on size structure of holm oak stands in Spanish dehesas: Implications for conservation and restoration. Environmental Conservation. 2003;**30**:61-70

[4] Diaz M, Campos P, Pulido FJ. The Spanish dehesas: A diversity in land-use and wildlife. In: Pain DJ, Pienkowski MW, editors. Farming and Birds in Europe: The Common Agricultural Policy and its Implication for Bird Conservation. London: Academic Press; 1997. pp. 178-209

[5] Montserrat P, Fillat F. The systems of grassland management in Spain. Managed grasslands. 1990;**17**:37-70

[6] López Sáez JA, López GP, López ML, Cerrillo CE, González CA, Prada GA. Origen prehistórico de la dehesa en Extremadura: Una perspectiva paleoambiental. Revista de Estudios Extremeños. 2007;**63**:493-510

[7] Clemente Ramos J. Ocupación del espacio y formas de explotación. Dehesas y adehesamiento en Extremadura (c. 1250 - c. 1450). In: García Fitz F, Jiménez Alcázar JF, editors. La historia peninsular en los espacios de frontera: Las "Extremaduras históricas" y la"Transierra" (Siglos XI-XV). Murcia: Sociedad Española de Estudios Medievales, Editum; 2012. pp. 253-277

[8] Llorente Pinto JM. El problema de la sostenibilidad de las dehesas a la luz de la evolución histórica de los terrenos adehesados. Cuadernos de la Sociedad Española de Ciencias Forestales. 2003;**16**:135-140

[9] Font Tullot I. Climatología de España y Portugal. Salamanca: Ediciones Universidad de Salamanca; 2007. 425 p

[10] Maestre F, Ramírez D, Cortina J. Ecología del esparto (Stipa tenacissima L.) y los espartales de la Península Ibérica. Revista Ecosistemas. 2007;**16**:111-130

[11] Pôças I, Cunha M, Pereira LS. Pastagens seminaturais de montanha: Lameiros, sistemas ancestrais no século XXI. Taller CYTED XVII, El agua en Ibero-América: Tecnologías apropiadas y tecnologías ancestrales. Lima: Universidad Nacional de Piura-Perú; 2006. 25 p

[12] González UE. Industrialización y desarrollo metropolitano en España. Ería. 1991;**26**: 199-215

[13] Shucksmith M, Thomson KJ, Roberts D, editors. The CAP and the Regions: The Territorial Impact of the Common Agricultural Policy. Wallingford: CABI Publishing; 2005. 247 p

[14] IGME. Mapa Geológico de la Península Ibérica, Baleares y Canarias a escala 1:1.000.000. Madrid: Instituto Geológico y Minero de España; 1995

[15] IGN. Sistema de Información sobre Ocupación del Suelo en España (SIOSE). Madrid: Instituto Geográfico Nacional; 2011

[16] CLC. Corine Land Cover. European Environment Agency; 2012.

[17] Clary J. Rainfall seasonality determines annual/perennial grass balance in vegetation of Mediterranean Iberian. Plant Ecology. 2008;**195**:13-20

[18] Kottek M, Grieser J, Beck C, Rudolf B, Rubel F. World map of the Köppen-Geiger climate classification updated. Meteorologische Zeitschrift. 2006;**15**:259-263

[19] Feo PF. La deforestación en la provincia de Madrid como consecuencia del proceso desamortizador. Estudios Geográficos. 1984;**177**:475-482

[20] Olea L. San Miguel-Ayanz a. The Spanish dehesa. A traditional Mediterranean silvopastoral system. In: Lloveras J, González-Rodríguez A, Vázquez-Yáñez O, Piñeiro J, Santamaría O,

Olea L, Poblaciones MJ, editors. Sustainable Grassland Productivity. Proceedings of the 21st General Meeting of the European Grassland Federation. Badajoz: Organizing Committee of the 21st General Meeting of the European Grassland Federation, Sociedad Española para El Estudio de Los Pastos. 2006. pp. 3-13

[21] Guerra CA, Pinto-Correia T. Linking farm management and ecosystem service provision: Challenges and opportunities for soil erosion prevention in Mediterranean silvopastoral systems. Land Use Policy. 2016;**51**:54-65

[22] Scholes RJ, Archer SR. Tree-grass interactions in savannas. Annual Review of Ecology and Systematics. 1997;**28**:517-544

[23] Manning AD, Fischer J, Lindenmayer DB. Scattered trees are keystone structures - implications for conservation. Biological Conservation. 2006;**132**:311-321

[24] Moreno G, Pulido FJ. The functioning, management and persistence of dehesas. In: Rigueiro-Rodríguez A, McAdam J, Mosquera-Losada MR, editors. Agroforestry in Europe: Current Status and Future Prospects. The Netherlands: Springer; 2009. pp. 127-160

[25] Joffre R, Vacher J, de los Llanos C, Long G. The dehesa: An agrosilvopastoral system of the Mediterranean region with special reference to the sierra Morena of Spain. Agroforestry Systems. 1988;**6**:71-96

[26] Gómez-Sanz V, Bunce RGH, Elena-Roselló R. Landscape assessment and monitoring. In: Azevedo JC, Perera AH, Pinto MA, editors. Forest Landscapes and Global Change: Challenges for Research and Management. New York: Springer; 2014. pp. 199-226

[27] Lasanta-Martínez T, Vicente-Serrano SM, Cuadrat-Prats JM. Mountain Mediterranean landscape evolution caused by the abandonment of traditional primary activities: A study of the Spanish central Pyrenees. Applied Geography. 2005;**25**:47-65

[28] Rescia AJ, Pons A, Lomba I, Esteban C, Dover JW. Reformulating the social-ecological system in a cultural rural mountain landscape in the Picos de Europa region (northern Spain). Landscape and Urban Planning. 2008;**88**:23-33

[29] Calvo-Iglesias MS, Crecente-Maseda R, Fra-Paleo U. Exploring farmer's knowledge as a source of information on past and present cultural landscapes: A case study from NW Spain. Landscape and Urban Planning. 2006;**78**:334-343

[30] Biurrun I, García-Mijangos I, Berastegi A, Ambarli D, Dembicz I, Filibeck G, Jandt U, Janišová M, Jaunatre R, Kącki Z, Kuzemko AA, Pedashenko H, Polchaninova NY, Dengler J. Diversity of dry grasslands in Navarre (Spain): Experiences from the 7th EDGG field. Workshop, 15-24 June 2014. Bulletin of the European Dry Grassland Group. 2014;**24/25**:4-21

[31] Bascones-Carretero JC. Los pastizales de la Navarra húmeda. In: Publicaciones de Biología de la Universidad de Navarra, Serie Botánica, 1. Pamplona: Servicio de Publicaciones de la Universidad de Navarra; 1982. pp. 61-85

[32] Berastegi A, Peralta J, Lorda M, Remón J, García-Mijangos I, Biurrun I. Listado de los tipos de pastizales y prados presentes en Navarra. In: Canals Treserras RM, Emeterio

Garciandía L, editors. Nuevos retos de la ganadería extensiva: Un agente de conservación en peligro de extinción. Pamplona: Sociedad Española para el Estudio de los Pastos; 2012. pp. 240-245

[33] Castaños Ugarte PM. El pastoreo y la ganadería durante la romanización en el País Vasco. Isturitz. 1997;**9**:659-668

[34] Mangado Urdániz JM, Barbería MA, Oiarbide MJ. Pastoreo de ovino en ecológico en los secanos semiáridos de la Ribera de Navarra. Navarra Agraria. 2008;**171**:39-46

[35] Echeverría T, Asarta A. El ganado vacuno de raza pirenaica en Navarra. Pastos. 2011; **6**:213-234

[36] Mucina L, Bültmann H, Dierßen K, Theurillat JP, Raus T, Čarni A, Šumberová K, Willner W, Dengler J, García RG. Vegetation of Europe: Hierarchical floristic classification system of vascular plant, bryophyte, lichen, and algal communities. Applied Vegetation Science. 2016;**19**:3-264

[37] Rivas MS, Gandullo J, Serrada R, Allué J, Montero J, González J. Mapa de series de vegetación de España y memoria. Madrid: Ministerio de Agricultura, Pesca y Alimentación, ICONA; 1987

[38] Mac Donald D, Crabtree JR, Wiesinger G, Dax T, Stamou N, Fleury P, Gutierrez Lazpita J, Gibon A. Agricultural abandonment in mountain areas of Europe: Environmental consequences and policy response. Journal of Environmental Management. 2000;**59**:47-69

[39] Allende ÁF, López EN, Fernández GF. Caracterización fitoclimática de un sector de la montaña central cantábrica: el macizo de Curavacas (Palencia). Estudios Geográficos. 2015; **76**:7-38

[40] Sainz Ollero H, Moreno Saiz JC. Flora vascular endémica española. In: La diversidad biológica de España. Madrid: Pearson Education; 2002. pp. 175-195

[41] Montiel Molina C. Tradición, renovación e innovación de los usos y aprovechamientos en las áreas rurales de montaña. Cuadernos Geográficos. 2003;**33**:7-26

[42] Cortina J, Maestre FT, Ramírez D. Innovations in semiarid restoration. The case of Stipa Tenacissima L. steppes. In: Bautista S, Aronson S, Vallejo RJ, editors. Land Restoration to Combat Desertification. Innovative Approaches, Quality Control and Project Evaluation. Valencia: Fundación CEAM; 2009. pp. 121-144

[43] Pulido M, Schnabel S, Lavado Contador JF, Lozano-Parra J, González F. The impact of heavy grazing on soil quality and pasture production in rangelands of SW Spain. Land Degradation and Development. 2016. DOI: 10.1002/ldr.2501

[44] Correal E, Erena M, Ríos S, Robledo A, Vicente M. Agroforestry systems in southeastern Spain. In: Rigueiro-Rodríguez A, McAdam J, Mosquera-Losada MR, editors. Agroforestry in Europe: Current Status and Future Prospects. Dordrecht: Springer Netherlands; 2009. pp. 183-210

[45] Carrión JS, Fernández S, Jiménez-Moreno G, Fauquette S, Gil-Romera G, González-Sampériz P, Finlayson C. The historical origins of aridity and vegetation degradation in southeastern Spain. Journal of Arid Environments. 2010;**74**:731-736

[46] López OI. Entre la tradición y el cambio: La respuesta de la Región de Murcia a la crisis de la agricultura tradicional. Historia Agraria: Revista de Agricultura e Historia Rural. 1999;**19**:75-113

[47] Romero J, Francés M, editors. La huerta de Valencia: Un paisaje cultural con futuro incierto. Valencia: Universitat de València; 2012. 208 p.

[48] Van Doorn AM, Correia TP. Differences in land cover interpretation in landscapes rich in cover gradients: Reflections based on the montado of South Portugal. Agroforestry Systems. 2007;**70**:169-183

[49] Vieira LM, Eden P. The Portuguese Montados. In: McCracken DI, Bignal EM, Wenlock SE, editors. Farming on the Edge: The Nature of Traditional Farmland in Europe. Peterborough: Joint Nature Conservation Committee; 1995. pp. 99-102

[50] Gea-Izquierdo G, Cañellas I, Montero G. Acorn production in Spanish holm oak woodlands. Investigación Agraria: Sistemas y Recursos Forestales. 2006;**15**:339-354

[51] Oliveira F, Moreno G, López L, Cunha M. Origem, distribuição e funções dos sistemas agro-florestais. Pastagens e forragens. 2007;**28**:93-115

[52] Bachoud A, Dreyfus-Armand G, Posty PM. Exils et migrations ibériques au XXe siècle. Paris: Centre d'Etudes et de Recherches inter-européennes contemporaines (CERIC); 1996

[53] Moradiellos E. La España de Franco, 1939-1975: Política y sociedad. Madrid: Editorial Síntesis; 2000. 320 p

[54] Ortega CN. El proceso de mecanización y adaptación tecnológica del espacio agrario español. Madrid: Ministerio de Agricultura, Pesca y Alimentación; 1983

[55] Martínez Ruiz E. La España de Franco (1939-1975). Economía. Revista de Historia Económica-Journal of Iberian and Latin American Economic History. 2003;**21**:199-212

[56] Groome HJ. El desarrollo de la política forestal en el Estado Español: desde la Guerra Civil hasta la actualidad. Arbor. 1988;**505**:65-110

[57] Alonso GJ. España 1940-1960: Crecimiento económico. Revista de Estudios Agrosociales. 1982;**121**:81-125

[58] Simpson J, Serrano M. La agricultura española (1765-1965): La larga siesta. Madrid: Alianza; 1997. 416 p

[59] Acosta NR. Las dehesas de la sobremodernidad. La cadencia y el vértigo. Badajoz: Diputación Provincial de Badajoz; 2008. 508 p.

[60] Campos Palacín P. La degradación de los recursos naturales de la dehesa. Análisis de un modelo de dehesa tradicional. Agricultura y Sociedad. 1983;**26**:289-380

[61] Gonzalo Langa J. El impacto de la aplicación de la PAC en las producciones ganaderas de la Dehesa (1986-2010). In: Coleto Martínez JM, De Muslera Pardo E, González Blanco R, Pulido García F, editors. La agricultura y la ganadería extremeñas: Informe 2010. Badajoz: Caja de Ahorros de Badajoz; 2011. pp. 181-196

[62] Lavado Contador JF, Pulido FM, Schnabel S, Herguido SE. Fragmentation of SW Iberian rangeland farms as assessed from fencing and changes in livestock management. Effects on soil degradation. In: Alphan H., Atik M., Baylan E., Karadeniz N., editors. Proceedings of the International Congress on Landscape Ecology, 23-25 October 2014, Antalya, Turkey. Antalya: PAD Publications No: 2; 2015. pp. 183-192.

[63] Gaspar P, Escribano M, Mesías FJ, de Ledesma AR, Pulido F. Sheep farms in the Spanish rangelands (dehesas): Typologies according to livestock management and economic indicators. Small Ruminant Research. 2008;**74**:52-63

[64] Moreno G, Rolo V. Facilitación de la regeneración de Quercus ilex en dehesas ibéricas por dos especies contrastadas de matorral: jara y retama. Cuadernos de la Sociedad Española de Ciencias Forestales. 2014;**40**:159-166

[65] Herguido Sevillano E, Lavado Contador JF, Pulido M, Schnabel S. Spatial patterns of lost and remaining trees in the Iberian wooded rangelands. Applied Geography. 2017;**87**:170-183

[66] Bota G, editor. Ecology and Conservation of Steppe-Land Birds: International Symposium on Ecology and Conservation of Steppe-Land Birds. Lleida: Lynx Edicions; 2005

[67] Rico E. Política forestal y conflictividad social en el noroeste de España durante el primer franquismo, 1939-1959. Historia Social. 2000;**38**:117-140

[68] Madrigal A. Problemática de la ordenación de masas artificiales en España. Cuadernos de la Sociedad Española de Ciencias Forestales. 1998;**6**:13-20

[69] Soto Caba MÁ. El consumo de papel y sus implicaciones sobre los bosques y el medioambiente. Cuadernos de Biodiversidad. 2005;**17**:21-26

Chapter 3

Vegetation Changes in the Miombo Woodlands in Northwestern Zimbabwe: A Case Study of Nkayi District 1990 to 2017

Albert Chirima, Peter Mundy, Nicholas Ncube and Andre F. VanRooyen

Additional information is available at the end of the chapter

http://dx.doi.org/10.5772/intechopen.72466

Abstract

The research applied Geographic Information Systems (GISs) and remote sensing tools in quantifying land cover changes in Nkayi District and assess the drivers for such changes. This was done to link the impacts of anthropogenic activities to change in the physical environment especially looking at ecosystem goods and services, which in turn reduce their productivity. Satellite images were analyzed for 1990, 2000, 2010, and 2017 in order to produce temporal land cover maps for Nkayi District and use them as tools for estimating the rates and the extent to which land cover has changed from 1990 to 2017. Four main land cover types were identified, namely woodland, deforested land, cultivated land, and water bodies. In 1990, woodland covered 58% of the total land area in Nkayi District, while deforested land, cultivated land, and water bodies covered 31, 11, and 0.2%, respectively. From 1990 to 2017, woodland declined to 47% in 2017, while deforested land and cultivated land increased to 14.9 and 36%, respectively. The major drivers of land cover changes were increase in household numbers, which were associated with woodland clearing for agriculture. The other drivers of land cover changes were soil infertility and overgrazing by livestock. The research was crucial in detecting the problems of forage shortages and poor rangeland conditions, mainly caused by expanding fields coupled with infertile Kalahari sands. The research highlighted the urgent need to manage the fragile miombo woodlands, which are being threatened by the increased demand for land for human settlements and cultivation. Alternatively, the research also highlights the need for farmers to produce more biomass in their fields in the form of high-value crop residues to cater for the loss of rangelands.

Keywords: miombo woodlands, vegetation, GIS, remote sensing, land cover change, participatory GIS

1. Introduction

Miombo woodland ecosystems are integral and crucial to biodiversity and human livelihoods for communities living across much of central and southern Africa stretching from Angola, Botswana, the Democratic Republic of Congo, Malawi, Mozambique, Namibia, South Africa, Tanzania, Zambia, and Zimbabwe. The woodlands are vast ecosystems consisting of *Brachystegia* species, *Baikiaea plurijuga, Combretum, Terminalia,* and acacia species [1, 4] and offer a lot of benefits to the local communities in the form of rangelands, firewood, traditional medicines, and the general esthetic of the environment [16]. They are also home to a diverse range of wildlife that includes antelopes, giraffes, rhinos, lions, and some of the largest populations of elephants in Africa [5]. However, over the years, these ecosystems have experienced vast changes in their physical nature, which affect the roles they play as a result of land degradation caused by humans who clear land for cultivation [3]. Land degradation manifests itself in many forms, and the common way is through the transformation of the land from its natural state into man-made land use forms, e.g. from forest into farmlands. Management of the transitions and the new land cover forms creates chances for land degradation to occur. The miombo woodlands are affected by three main interacting disturbance factors namely people, wild fires, and wildlife. Anthropogenic activities center on the partial to complete clearance of woodland, conversion of woodland to cropland, excessive resource extraction and overgrazing, which lead to the modification or transformation of the ecosystem. Land cover change is a major driving force of habitat modification and has important implications for the distribution of savanna vegetation, ecology, and biodiversity [6–8]. Abrupt land cover change is continuously changing ecosystems, thus threatening sustainability of wildlife species, functionality of ecological system, and livelihoods [6, 7, 10]. This is the case for most communal areas in the semiarid areas of Zimbabwe.

Land use change is often the first consequence of population and economic growth, and it can be argued that low soil fertility, lack of infrastructure, and the presence of diseases have been the main factors preserving miombo [3, 7, 19]. However, due to increased demand for land for cultivation, a new dimension was created leading to massive land cover changes and degradation. In Zimbabwe, the demand for land can be traced back to the period of the 1950s and since then, the population has continuously increased, which in turn led to the clearing of land for farming and grazing of livestock.

These human activities have been happening in fragile miombo environments where the annual rainfall is below average (<600 mm per annum) and the soils are inherently infertile. The impact and extent of human activities on land in the miombo woodlands are not fully understood and have not been clearly investigated. This research, therefore, aims at using satellite images to quantify the land cover changes from 1990 to 2017 and through participatory GIS seeks to determine the drivers of these changes.

Vegetation cover change, change in species composition, and plant quality are often viewed as indicators of susceptibility to land degradation and its severity [17, 22, 27]. Causes of the land cover changes range from population increase, overgrazing, fuel wood extraction, timber extraction to wood cutting for the purposes of building [2, 5, 9, 10, 19]. Campbell and Mapaure [5] estimated the demand for firewood and calculated that 9,285,000 tonnes of wood are burnt in Zimbabwe every year, with an estimated 7,141,000 tonnes of wood consumed in communal areas only, representing 77% of the total firewood consumed in 1994. Campbell and Mapaure [5] also noted that the demand for firewood has been increasing over the period 1990–1996.

Several sensor systems operating in the visible and near-infrared (NIR) currently operational include the Landsat series, SPOT, AVHRR, and MODIS and these are extremely useful particularly for vegetation mapping [21]. The selection of the sensor data depends on a number of factors including spatial and temporal resolution as well as the cost of securing the data [18, 26]. For example, AVHRR data have a pixel resolution of 1 km × 1 km, which can be classified as coarse, but the images have a much higher re-visitation period that makes them useful in landscape-scale vegetation phenology studies [25–27]. Sabins [21] and Weng [26] used high-resolution Landsat satellite sensors for land cover characterization in different environments, and thus, they are widely adopted for that purpose. The very high spatial resolution images provide fine details on the ground and can be used as ground truth data, for example, Quick Bird images, IKONOS, and Hyperion data [14]. However, the cost of these images is prohibitive to cover large areas of study or for multi-temporal analysis [24]. As a result, Landsat images provide a balance between spatial resolution, temporal resolution, and cost, which make them popular with researchers. The application of remote sensing, with frequent repetitive image acquisition in multiple wavelengths and Geographic Information System (GIS) technologies, has been used as powerful tools to investigate, detect, and analyze habitat change [12, 21, 24, 26], where vegetation indices are the capstone of the field. Satellite imagery has, over the years, been used for land cover classification, land use monitoring, and assessment [19, 22, 26], through providing information on land features, their condition and distribution. Equally, it has proven to be valuable in qualitative and quantitative terrestrial land cover changes. Furthermore, applied in rangelands, remote sensing has the advantage of identifying early land degradation risks [13]. Its application is therefore essential in developing scientific principles that result in improved sound conclusions. GIS and remote sensing tools are used in conjunction with PGIS to understand the changes to the miombo woodlands as a result of increasing human and livestock populations.

1.1. Description of study area

The study was carried out in Nkayi District in Matabeleland North Province of Zimbabwe (see **Figure 1**). Nkayi District is located in Natural Region IV of Zimbabwe, which is characterized by low rainfall, high temperatures in the summer reaching up to 35°C, and infertile sandy soils. Rainfall ranges from 350 to 940 mm and averages 591 mm per annum. The district is subject to seasonal drought and long dry spells during the rainy season, which makes it unsuitable for rain-fed agriculture. The area can, however, support extensive livestock production. The natural vegetation of Nkayi District is typical dry Savanna dominated by the miombo woodland. Tree species are mainly *Brachystegia* species, *Baikiaea plurijuga, Julbernardia globiflora, Combretum* and *Terminalia* species, acacia species, *Peltophorum africanum, Albizia amara, Albizia harveyi, Euclea divinorum, Grewia* species, and *Kirkia acuminata.* The common grasses in natural woodlands include *Aristida* species, *Eragrostis* species, *Digitaria* species, *Cenchrus ciliaris, Enneapogon cenchroides, Pogonarthria squarrosa, Schmidtia pappophoroides,* and *Stipagrostis uniplumis* [7, 8].

1.2. Materials and methods

1.2.1. Satellite image preparation and pre-processing

A time series analysis of Landsat TM and ETM images was done in order to deduce land cover changes over time. Orthorectified satellite images path 171 row 73 covering region of study area

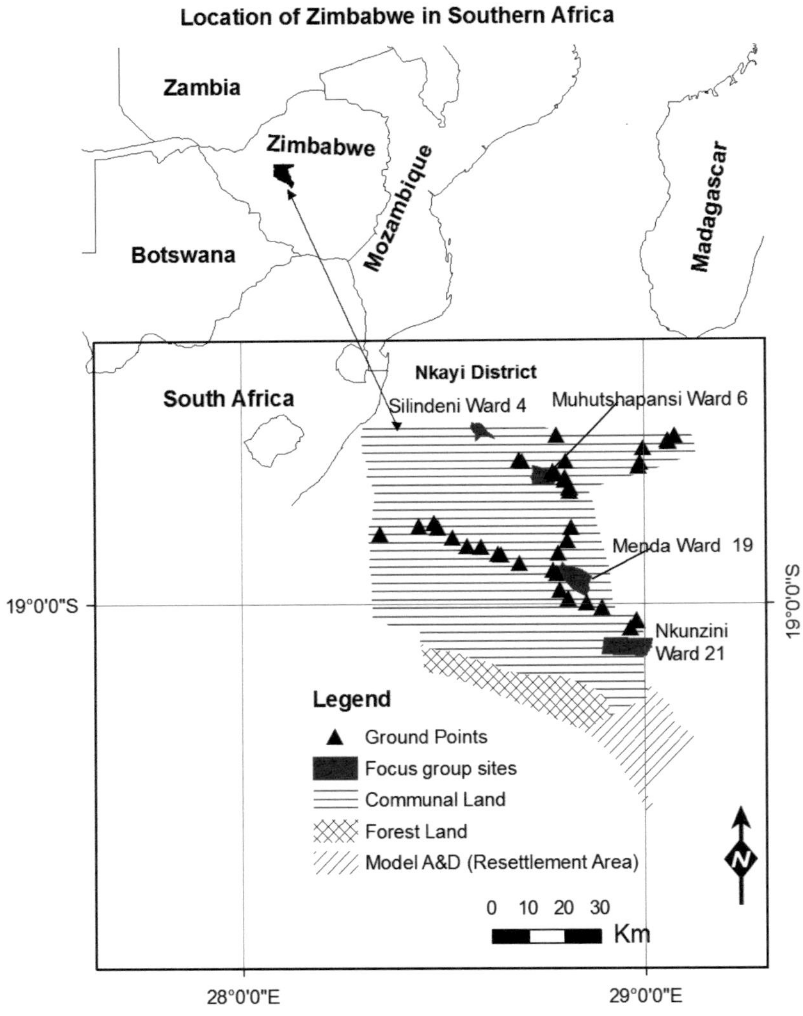

Figure 1. Location of Zimbabwe in southern Africa showing ground truth points in Nkayi District in Zimbabwe where the study was conducted.

(Nkayi District) were acquired from NASA, https://dds.cr.usgs.gov/. The satellite images were selected for the period toward the end of the vegetative growth from March up to beginning of May. This was done so that the contrast between vegetated areas and non-vegetated areas is clearly separable. The selected images had zero percent cloud cover, and were corrected for any geometric anomalies. The image for 2017 was georeferenced using Ground Control Points (GCPs) captured using Garmin Etrex 30, and the images for 2010, 2000, and 1990 were co-registered to the 2017 image in order to facilitate timeline comparison of land covers. In order to improve on the quality of the satellite images, the Digital Numbers (DNs) were converted to reflectance through a calibration process called normalization where Digital satellite numbers are converted to radiance and then radiance values are converted to sensor reflectance. This was done in order to correct for sun angle, sensor angle, and sensor height among other geometric distortions. The following formulas were used: $DN\ to\ Radiance = (Lmax - LMin)/255 * DN + LMIN$, and $Radiance\ to\ Reflectence = (d\pi * L\lambda * DN)/ESUN\lambda * Cos\theta$.

where ρ is the planetary reflectance, L_λ is the spectral radiance at sensor's aperture, $ESUN_\lambda$ is the band-dependent mean solar exoatmospheric irradiance, θ is the solar zenith angle, and D is the earth-sun distance, in astronomical units.

1.2.2. Image classification

We used Envi 5.1 image analysis software to prepare and analyze the satellite images. Training data also known as end members in Spectral Angle Mapper used in the supervised classification technique were generated in the field using a GPS and then the Region of Interest polygons for each representative land cover was digitized on the image using the Regions of Interest (ROI) tool. The regions of interest that were digitized were woodlands, cultivated lands, disturbed woodlands which also included fallow lands, as well as water bodies. A supervised classification technique was employed to classify the land covers for each year. The classification was done using a combination of unsupervised K-Means, visual analysis, and the Spectral Angle Mapper algorithm. K-Means unsupervised classification was used to generate classes showing the general view of the area, which aided in training site selection. A supervised classification technique was applied in the form of the Spectral Angle Mapper (SAM), which is a physically based spectral classification technique that uses an n-dimensional angle to match pixels to reference spectra. This technique was used on calibrated reflectance data which is known to be relatively insensitive to illumination and albedo effects. The spectra for different classes were compared to typical spectra of known land features for the purpose of separating ambiguity features. In semiarid areas like Nkayi, it is difficult to use texture of the image because the land covers are always a mixture of different plant and grass species with distinct soil cover. At this point, visual interpretation techniques and prior knowledge were also employed to determine the features based on characteristics such as site, location, arrangement, and their shape.

1.2.3. Change detection analysis

After the supervised classification process, a land cover change matrix was calculated to show the direction of the changes. The land cover change matrix was calculated using the 1990 classified image as the input and the 2017 classified image as the output. The Change detection algorithm was used in Envi 5.1.

1.2.4. Participatory GIS analysis

The observed land cover changes were verified by the local communities through participatory GIS, as well as explaining the direction and causes of the land cover changes. The participatory GIS was carried out to verify the results of satellite image classification and also to explain the changes observed on satellite images. A stratified random sample was used to select four wards with two wards selected in the north while two wards were selected in the southern part of the district. In each ward, one village was selected for focus group discussions. On average, 35 participants were randomly selected from each village to participate in focus groups. Participants drew village resource maps for 1990 and for 2017 indicating the location of key resources within the villages. Plenary sessions were held during which the results from resource mapping and change pattern were further scrutinized and validated. The data sets from all the four focus groups were then collated, coded, and analyzed through use of Kenda matrices.

1.2.5. Accuracy assessment

The Kappa statistic and Google Earth Pro were used to assess the accuracy of the classification. The confusion matrix calculated from sampled points yielded an 88% Kappa while those sampled using Google Earth yielded a Kappa statistic of 91%.

2. Results

2.1. Spatial distribution of land covers and land cover change from 1990 to 2017

Land cover changes in Nkayi District were analyzed through the interpretation of Landsat TM and ETM images for 1990, 2000, 2010, and 2017. From the image classification process, four main land cover classes were identified. The classes were woodland, cultivated land, deforested/degraded land, and water bodies (**Figure 2**). Although the spatial extent of sand beds was limited, they were delineated by their high reflectance values. The spatial distribution of these land covers for each year is highlighted in **Figure 2**.

2.1.1. Distribution of miombo woodlands

Wooded areas were defined as a range of dense woody trees mixed with grass species. Woodland areas in Nkayi District were divided into three categories comprising the demarcated national forest land forming the Gwampa forest on the southwestern part of the district, the communally managed woodlands and those in the resettlement areas formerly known as commercial farms before the land reform program. Gwampa Demarcated Forest is managed

http://dx.doi.org/10.5772/intechopen.72466

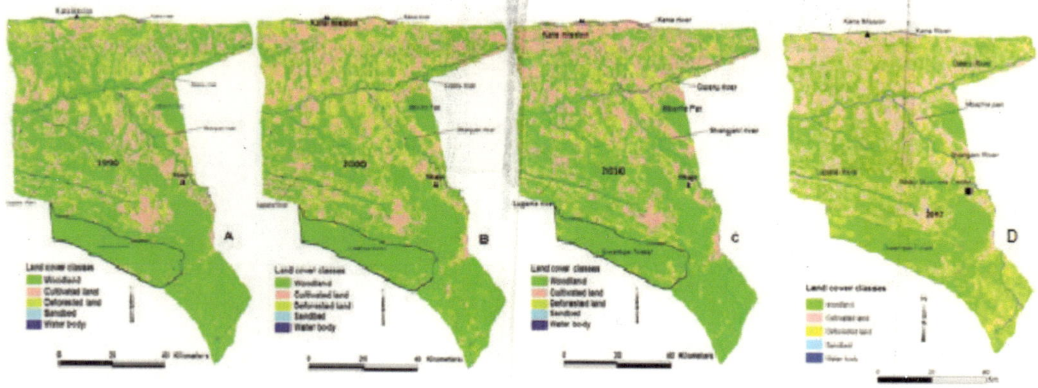

Figure 2. Distribution of land covers in Nkayi District from 1990 to 2017.

by the Forestry Commission under the Forestry Commission Act of 1996. Also found in the southern part of the district were resettlement farms. These were formerly commercial farms that were resettled to small-scale farms under the land reform models A2 and D. The rest of the district was occupied by communal woodlands stretching from the middle of the district up to the northern part of the district. Here, the fragmented woody cover of mainly *Brachystegia spp.* woodlands in uplands and mixed *Brachystegia spp.* and acacia toward the valleys is dominant. Woodlands were the dominant land cover by area/size in the whole of Nkayi District, covering 58% in 1990 before dropping to 50% and subsequently 47% in 2017 (**Table 1**). Communal areas had a very low proportion of woodlands comprising only 54% of the land area in 1990 before dropping to 44% in 2017. Land allocated to cultivation activities in communal areas was very high, 42% in 1990, and continues to increase as the demand increased. In contrast, woodlands in the resettlement areas occupied above 70% of the total land area in the same year. This indicates that the majority of the land in communal areas is allocated to cultivation (42%)-related activities, whereas in the resettlement areas, the majority of the land is woodland.

2.1.2. Spatial distribution of cultivated land

A large proportion of the land in Nkayi District is used for crop production during the rainy season between November and April and then becomes open for grazing after harvest time, between July and October. Livestock grazing is communal-based where unrestricted movement is permitted during the dry season. The total land under cultivation in the whole district in 1990 was 31% and increased to 36% in 2017 (**Table 1**). The majority (98%) of the cultivated lands were located in the communal areas, whereas only 2% were found in the resettlement area. Field observations showed that a very large proportion of these fields were on sandy soils, while in contrast, the proportion of the cultivated lands in the resettled areas were on clay soils that were highly fertile and suitable for crop production.

Land cover type	Total land cover (000 ha)				% of Land cover			
	1990	2000	2010	2017	1990	2000	2010	2017
Woodland	306	257	262	252	57.8	49.5	50	47
Cultivated land	164	167	186	193	31	31.8	35	36
Deforested land	58	97	81	84	11	18.5	14.9	16
Water	1	0.7	0.5	2	0.2	0.13	0.1	0.3

Table 1. Land cover change in Nkayi District: 1990-2017.

2.1.3. Distribution of deforested land

The deforested land consisted of open barren land, sparsely distributed woody and grass species, and abandoned fields which had not fully transitioned back to woodland. Most deforested lands were found close to cultivated lands in the form of fallow lands and also near homesteads where livestock graze during the planting season. Deforestation was also associated with areas around water points, as a result of trampling by livestock. During plowing, oxen used for ox-drawn plows graze in these areas. Together with fields and woodlands, deforested lands provided forage for the rest of the livestock and formed an important source of livestock feed. Since they were mainly close to homesteads, they were used as temporary grazing areas during the times of emergencies, as such were a crucial forage source for livestock. Up to 11% of the land was occupied by deforested lands in 1990 (**Table 1**) which translated to 58,000 hectares of land rising to 84,000 hectares in 2017. Of this, about 68% were found in the communal areas, while 32% were found in the resettled areas.

2.1.4. Water bodies

Water bodies such as dams, open wells, and boreholes were very scarce yet so vital in the district. Although they occupy much less than 1% (**Table 1**) of the land area in Nkayi, their availability and distribution play an important role in making optimal use of the land. They supply drinking water for domestic and livestock purposes and also used for irrigation services [11]. Three of the four surveyed villages had access to an open water source located within their villages. Access to water sources is a big challenge in the district, as witnessed by less than 1% of water sources available and there is huge competition for water between humans and livestock. This has limited the development of agricultural activities in the district. Agricultural extension officers also acknowledged that the scarcity of water sources in the district hindered the development of irrigation schemes. They stated that there were less than five operational irrigation schemes in the whole district. However, they revealed that small individual gardens were available along the major rivers such as Gweru, Shangani, and Kana rivers, and these were only enough to serve small local communities.

2.2. Spatial distribution of vegetation condition

Function and health of the miombo woodlands is critical for their survival and the life they support. Normalized Difference Vegetation Index (NDVI) for the district was calculated for 2017 to assess the distribution of vegetation healthiness across the districts. **Figure 3** shows the distribution of vegetation vigor across Nkayi District. High shades of red (0.4–0.62) depict healthy and dense vegetation, which were mainly woodlands inside the Gwampa forest and resettlement areas. Green shades represented sparse vegetation and grasslands with signs of degradation represented in yellow shades while severely degraded areas were represented in shades of light blue to light green and these marked the transitions from woodland to degraded land and from cultivated land to degraded land.

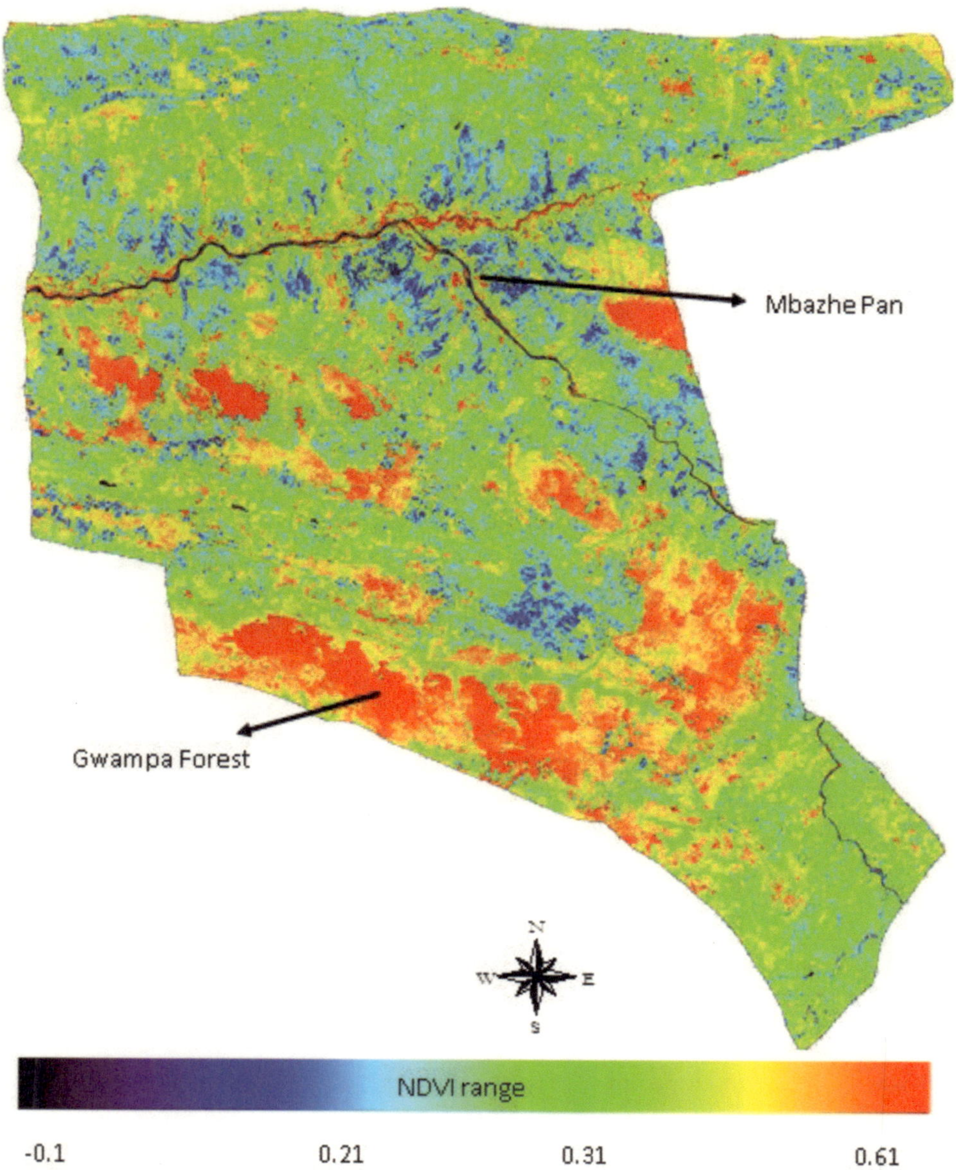

Figure 3. NDVI image showing degraded land in varying shades of blue while healthy woodland is shown in shades of red.

2.3. Magnitude and rate of land cover changes

The study showed widespread changes in land cover in Nkayi District during the last two and a half decades. This was evidenced by the widespread decrease of woodland in the communal areas and a visible increase of cultivated and deforested lands as shown in **Tables 1** and **2**.

Woodlands were decreasing at a rate of 0.76% per annum, translating to an average loss of 2300 ha per annum. On the other hand, cultivated land and deforested lands were increasing at the rates of 1.3 and 1.8% per annum, respectively (**Table 2**), which meant that they were increasing at 1800 and 800 ha per annum respectively. Considering the decrease in woodland against the increase in cultivated land and deforested land, it was clear that woodland declined because of expansion of cultivated land. There were also observed changes of deforested/degraded land being replaced by cultivated land especially between 2000 and 2010 images. This can be explained by the intricate shifting of cultivation by farmers whereby some lands left as fallow for some time were reclaimed after the ones being cultivated lost fertility, or as a result of inheritance of fallow lands by own children or relatives. Besides showing a general decrease in degraded/deforested lands, the overall trend showed that there were fluctuations in their extent from time to time.

2.4. Land cover change matrix 1990: 2017: Direction of change

Almost all the land cover classes went through some changes, with deforested and cultivated lands showing huge likelihood to changes. Twenty-two percent of the deforested lands remained unchanged between 1990 and 2017. Woodlands had the least likelihood to change as witnessed by 66% remaining unchanged between 1990 and 2017. Cultivated lands and water bodies had 55 and 40% of the areas unchanged. Only wooded areas showed a general decline of 11% in land covers during the study period, while cultivated lands, deforested land, and water bodies all showed an increase during the same period with deforested lands leading at 25% gain. Cultivated lands gained 31 and 13%, respectively, from woodlands and deforested land, while during the same period, fields lost 21 and 41% to woodlands and deforested areas,

Change detection statistics as percentages						
Land cover type	**Water**	**Forest**	**Deforested**	**field**	**Row total**	**Class total**
Water	**40**	0	1	0	98	100
Forest	28	**66**	28	31	100	100
Deforested	4	13	**22**	13	100	100
Field	27	21	49	**55**	100	100
Class total	100	100	100	100	0	0
Class changes	60	34	78	45	0	0
Image difference	87	−11	25	12	0	0

Table 2. Trade-offs among forest, cultivated land, field, and water bodies depicting direction of land change between 1990 and 2017.

respectively. Deforested land gained most of its land from cultivated lands in the form of fallowing, which was very common in the district. Regeneration was slow as witnessed by only 13% of deforested lands in 1990 detected as woodlands in 2017.

3. Discussion and conclusions

3.1. Understanding the direction of land cover changes

Analysis of satellite images revealed that there are complex and widespread land cover changes in Nkayi over the two and a half decades from 1990 to 2017. In general, there has been an increase in area under croplands and deforested lands. The increasing cultivated lands were mainly as a result of the conversion of woodland into croplands. As a result, 31% of woodlands were consumed by field lands since 1990. On the other hand, due to soil fertility challenges, some fields were being abandoned and left to regenerate back to woodlands. This resulted in 21% of cultivated lands detected as woodlands in 2017. During the fallowing periods, fallow lands were also being reclaimed for cultivation. These movements suggest that increase in cropland was through an intermediate conversion of woodlands into croplands and then deforested lands. In the process, old fields are created when cultivated lands are left uncultivated for a time long enough to regenerate into some sort of disturbed lands. As the old fields grow older, tree and grass species also grow, which turn the old field into woodland. Some old fields are also reclaimed back to croplands after being left for a couple of years as follows. There is also a direct movement from woodland to deforested land which is caused by direct impacts of overgrazing and wood extraction. During the first 10 years, farmers cleared a lot of woodland land and the decrease is closely matching that of increasing deforested land and cultivated land for crop production.

3.2. Drivers of land cover changes

Land cover changes have been noted all over the communal areas in Nkayi District, and the causes of these changes range from increase in household numbers and associated woodland clearing for agriculture, overgrazing by livestock, and soil infertility. The main driver for the expansion of croplands into forests appears to be the increase in household numbers. This is in contrast with the findings by Kamusoko and Aniya [15], Mapedza et al. [19], and Sibanda et al. [23] who cited population increase as the main factor in expansion of croplands in nearby Gokwe. Population increase did not necessarily result in increase in field sizes per se, but rather a new homestead is mandatorily allocated a field which basically means that a new area is opened up in the woodlands. In Nkayi, it was, however, noted that population increase was not a major factor as the number of people did not increase much, but the bigger change was noticed in household numbers. Population figures from the Zimbabwe National Census for the period 1992–2002 showed that household numbers increased from 110,161 to 111,118 because of the positive balance between births and deaths as well as immigration. Information from focus group discussions also indicates that these new immigrant households regularly clear land for establishing new crop fields and this led to an upsurge in area under cultivated land.

The other driver of land cover change is the impact of livestock on the rangelands (both forest land and degraded land). A decrease in forest land is closely related to a reduction in available rangeland. Feed shortages were the major setback for farmers in Nkayi District and elsewhere in miombo woodlands as reported by Hamandawana et al. [9] and Homann et al. [11]. The study also noted the same link between livestock feed shortages and shrinking of woodlands with the total livestock population remaining constant over the past 19 years, this means that livestock density increase as rangelands decrease. Woodlands' contribution to livestock feed, however, needs to be ascertained as more often, the quality of the grass species found there is of poor nutritional value. Also, increasing lands that are deforested highlight the decreasing amount of health rangelands available, i.e. changes in plant species composition.

Soil infertility is also noted in farmer discussions as another contributing driver of land cover changes. When farmers had worked in their fields since the1950s and through generations of use the fields became infertile and needed huge investments in soil fertility management to boost production. Faced with this problem, farmer group discussions indicated that they either cleared virgin land or extended their fields into the grazing areas. In the process, they clear vast lands for cultivation and the process starts again. On the other hand, old fields are then left as fallow lands. These, according to farmers, are left for the future generation and would be utilized and managed by the families concerned. Fallow lands are important in mitigating the impacts of wind erosion. They allow the soils to re-gain nutrients and provide an open space for cattle grazing. They are also used to grow thatching grass, fencing and building poles and as reserve grazing for the household's livestock only. They remain private and as such are not open for the whole community to utilize.

There are also, other socioeconomic factors which limit the ability of farmers to cope with the changing socioeconomic conditions and in a way contributing to land cover changes. People are relatively poor and are facing a number of challenges including costs of school fees, medical expenses, sourcing farming inputs and food to an extent that they are left with little or no resources to invest in productivity-enhancing technologies. This is forcing farmers to continue practicing unsustainable farming.

The decrease in forest land means that more land is exposed to heat, raindrop impact and runoff which increase the incidence of erosion. It also means that by reducing vegetation density on degraded lands, the loose topsoil is washed away easily, thereby increasing its loss. Eroded soils are deposited into rivers and dams which reduce their holding capacity and this contributes to dry season water shortages in the district. Again, reducing vegetation cover results in increased runoff and less infiltration of rainwater which also means that the groundwater recharge system is reduced therefore causing early drying up of wells and boreholes.

3.3. Vegetation healthy distribution

Woodlands, shown in different shades of red (**Figure 2**), were concentrated in the southern conservation area (Gwampa Demarcated Forest), and fragmented in the communal areas in the northern parts including the Mbazhe Pan area near the eastern boundary of the district. Gwampa Demarcated Forest Reserve has considerable health vegetation (high shades of red) because the area is reserved and is closer to Lake Alice, which is popular for its waterfowl and hence became part of the protected area [20]. The Mbazhe Pan also boasts of healthy vegetation because it was originally designated as a bird sanctuary in terms of the Parks and Wildlife Act (1975) (revised 1996). It is endowed with natural beautiful scenery, wetland and woodland habitats, bird life, and other biodiversity of importance [20]. Since the image was acquired during the post-harvest period (June), cultivated land and degraded rangelands had little or no photosynthetic activity as most of the crop residues and grasses had dried up or been removed by livestock resulting in bare surfaces hence low NDVI values of less than 0.2. Communities that surround Mbazhe Pan Sanctuary pose the greatest threat to the wetland through livestock grazing, siltation, erosion, and damage to infrastructure. As a result, the fence around the pan had been vandalized. Part of the wall had been destroyed and water drains into the Shangani River, which is less than a kilometer away. Signs of degradation and poor vegetation health were noticeable on the boundaries of the pan (**Figure 3**).

3.4. Conclusions

Integration of remote sensing and PGIS opened a better understanding of the drivers of land covers as there was a better way of interpreting the results by comparing classified images and people's perceptions. The study, therefore, concluded that the process of expanding croplands comes at the expense of forest and grazing lands. In terms of grazing land, the area available for grazing has decreased even though the herd size remained constant over the same period. This reduction has placed a lot of pressure on the ever-decreasing rangelands. The potential of the grazing land to offer enough goods and services such as high-quality biomass has been reduced. As a result, natural endowments of the land should be supplemented with sustainable inputs that improve the productivity of the land. The expanding crop fields must compensate for the loss of rangelands by producing equal or more biomass for feed as the rangelands did. Feed technologies, such as using crop residues as feed, using dual sweet sorghum cultivars, and intercropping conventional crops with fodder [1, 3], increase the productivity of the crop system to augment feed for livestock. Studies in the Philippines showed increased benefits in terms of feed quality and yield per hectare for intercropped sorghum and forage legumes, and they produce improved quality and quantity of sorghum fodder [1]. If these technologies are carefully practiced in semi-arid regions, they can potentially free a lot of pressure on rangelands, by producing extra feed.

The study recommends that Sub-Saharan Africa needs to capture, store, and recycle land resources whenever possible to improve the production system of the land resources. Studies by United Nation Convention to Combat Desertification (UNCCD) in 2004 acknowledge that "Even though water and nutrients are scarce in dry lands, much is still wasted through runoff, unproductive livestock systems, and wasteful irrigation system". Fallow lands can present a good entry point in managing the rangelands. Fallow lands are reserved as private land as they are constantly linked to the occupiers as grazing reserves or fallow fields. In most cases, they are also found closer to homes and this would mean that they are easily accessible and manageable. In this case, it can be better if these are turned into feedlots with high palatable grass species such as *bana* grass grown. This might enhance the productivity of such lands and at the same time providing feeding for livestock.

The study also recommends that managing the herd size, age, and health is tantamount to reducing overgrazing. Sound livestock management practices such as timing the off-take of cattle economically improve the system in that an animal is moved out of the system through sale before it has eaten too much feed and/or is too old. In addition, money generated can be invested back in the system to improve the health of other animals, hence improving their productivity.

Author details

Albert Chirima[1]*, Peter Mundy[1], Nicholas Ncube[1] and Andre F. VanRooyen[2]

*Address all correspondence to: amirihc2@gmail.com

1 National University of Science and Technology (NUST), Bulawayo, Zimbabwe

2 International Crops Research Institute for the Semi-Arid Tropics (ICRISAT), Bulawayo, Zimbabwe

References

[1] Azraf-ul-Haq A, Riaz A, Maeem M, Tanveer A. Performance of forage sorghum intercropped with forage legumes under different planting patterns. Pakistan Journal of Botany. 2007;**39**(3):431-439

[2] Bai ZG, Dent DL. Global Assessment of Land Degradation and Improvement: Pilot Study in Kenya. ISRIC: Nairobi; 2006

[3] Blummel M, Samad M, Singh OP, Amede T. Opportunities and limitations of food-feed crops for livestock feeding and implications for livestock water productivity. The rangeland journal. 2009;**31**:207-212

[4] Campbell BM. editor. Woodlands & Welfare in Africa. CIFOR: The Miombo Transition; 1996

[5] Campbell BM, Mapaure I. Changes in miombo woodland cover in and around Sengwa wildlife research area, Zimbabwe, in relation to elephants and fire. African Journal of Ecology. 2002;**40**(3):212-219

[6] Duerksen C, Snyder C. Nature-Friendly Communities: Habitat Protection and Land Use Planning. Island Press: University of Michigan; 2005 412 p

[7] Jewa EKK, Dougilla AJ, Sallua SM, Jerome O'Connell B, Benton TG. Miombo woodland under threat: Consequences for tree diversity and carbon storage. Forest Ecology and Management. 2016;**361**:144-153

[8] Gonçalves FMP, Revermann R, Gomes AL, Aidar MPM, Finckh M, Juergens N. Tree species diversity and composition of Miombo woodlands in south-Central Angola: A Chrono-sequence of Forest recovery after shifting cultivation. International Journal of Forestry Research. 2017

[9] Hamandawana H, Nkumbwe M, Chanda R, Eckardt F. Population driven changes in land use in Zimbabwe's Gutu district of Masvingo province. Some lessons from recent history. Applied Geography. 2005;**25**:248-270

[10] Hartemink AE, Veldkamp T, Bai Z. Land cover change and soil fertility decline in tropical regions. Turkish Journal of Agriculture. 2008;**32**:195-213

[11] Homann S, Van Rooyen A, Moyo T, Nengomasha Z. Goat Production and Marketing: Report on Baseline Information for Semi-Arid Zimbabwe. Bulawayo: ICRISAT; 2007

[12] Janssen LF, Huurnaman GC. Principles of Remote Sensing. Enschede: ITC; 2001

[13] Jensen ME. Spatial modelling of rangeland potential vegetation environments. Journal of Range Management. 2001;**54**:528-536

[14] Ji CY, editor. Land degradation in central Asia: Revised final report for the Asian Development Bank, Central Asian Countries Initiative for Land Management Multicountry Partnership Framework Support Project. Asian Development Bank. 2008;

[15] Kamusoko C, Aniya M. Land use/cover change and landscape fragmentation analysis in the Bindura District, Zimbabwe. Land Degradation and Development. 2007;**18**:221-233

[16] Katerere Y, Guveya E, Muir K. Community forest management: Lessons from Zimbabwe. IIED. 1999

[17] Lambin EF, Geist HJ, Lepers E. Dynamics of land-use and land-cover change in tropical regions. Annual Review of Environment and Resources. 2003;**28**:205-241

[18] Lillesand TM, Keifer W. Remote Sensing and Image Interpretation. New York: John Wiley; 1994

[19] Mapedza E, Wright J, Fawcett R. An investigation of land cover changes in Mafungabusi forest, Zimbabwe using GIS, remote sensing and participatory mapping. Applied Geography. 2002;**23**:1-21

[20] Mundy PJ, Chiweshe NC, Couto JT. Waterfowl counts at Mbazhe pan. Honeyguide. 2000;**46**:148-150

[21] Sabins FF. Remote Sensing: Principles and interpretation. New York: Freeman; 1996

[22] Scoones I. Transforming soils: The dynamics of soil fertility management in Africa. In: Scoones I, editor. Dynamics and Diversity: Soil Fertility and Farming Livelihoods in Africa. London: Earthscan; 2001. pp. 1-44

[23] Sibanda A, Homann S, Van Rooyen A, Dimes J, Nkomboni D, Sisito G. Understanding community perceptions of land use changes in the rangelands. Experimental Agriculture. 2011;**47**:153-168

[24] Thankabail M, Lyon JG, Huete A, editors. Hyperspectral Remote Sensing of Vegetation. Boca Raton: CRC Press; 2012

[25] Vlek PLG, Le QB, Amene TL, editors. Land Decline in Land Rich Africa: A Creeping Disaster in the Making. Rome: CGIAR Science Council Secretariat; 2008. 55 p

[26] Weng Q. Advances in Environmental Remote Sensing: Sensors, Algorithms and Applications. Boca Raton: CRC Press; 2011

[27] Wessels KJ, Prince SD, Frost PE, Van Zyl D. Can human induced land degradation be distinguished from the effects of rainfall variability? A case study in South Africa. Journal of Arid Environments. 2004;**68**:271-297

Chapter 4

Neotropical Forests from their Emergence to the Future Scenario of Climatic Changes

Cristiane Dias e Sarmento and
Marcel Giovanni Costa Franca

Additional information is available at the end of the chapter

http://dx.doi.org/10.5772/intechopen.72608

Abstract

Neotropical forests (NFs) play a main role in delivering environmental services such as biodiversity conservation and C sink. At the same time, these are some of the most disturbed vegetations in the world, since they are under accelerated rates of suppression and recovery. Conserving the remaining NF and recovering degraded areas is then urgent, although it is not an easy task. Ecological traits are widely varied across NF, as well as their responses to anthropic intervention. Generally, two large groups are observed according to climatic traits: (a) rain forests (RFs), in regions with 6 months or more of precipitation during the year and (b) seasonal dry forests (SDFs), in drier regions. Such forest types show very distinct species composition, α- and β-diversities, as well as functional and biomass dynamics. In this chapter, we both highlight the main differences between RF and DF, from their origin to present-day distribution, species composition, taxonomic and functional diversities, and discuss the predictions for shifts in all these traits during the next decades. Although few certainties, NF potential for mitigation of atmospheric C increases is a consensus among researchers. We also speculate about possible interventions, with the aim of avoiding a drastic future scenario.

Keywords: Anthropocene, C sequestration, deforestation, global warming, greenhouse effect, Gondwana, Latin America, selective logging

1. Introduction

Forests play a main role in global ecosystem services as water supply, climate regulation, conservation of biodiversity richness, and carbon dynamic and storage. In this way, they are at local, regional, and global levels, one of the most important mitigators of the environmental changes that characterize the Anthropocene era. In spite of this, changes on forest cover and structure, most of them being deleterious alterations, are still being intensified throughout the world. For instance, 15.3 billion of the current 3.04 trillion living trees in the planet are vanishing annually [1], as a result, ~192,000 km^2 of forest area was lost per year, which means that 2.3 million km^2 of forest was depleted in the first dozen years of the present millennium. Unfortunately, numbers for global forest regrowth are more modest, with the regeneration of only 1.0 million km^2 in the same period. Deforestation impacts are still higher in the tropics, where forest loss is increasing in a rate of 2101 km^2 $year^{-1}$ [2]. As this vegetation stocks a massive amount of 55% of the C of global forests (471 Pg), such dynamics results in a release of approximately 3 Pg C $year^{-1}$ into the atmosphere [3], making their conservation an urgent issue. In this context, Neotropical forests (NFs) have a noteworthy function, as they hold almost 50% of the C pool in tropical forests [4] (**Figure 1**). Furthermore, these forests were although considered as a top priority for maintenance of global biodiversity [5], they are widely depleted for agriculture and livestock expansion in both Meso and South Americas.

Due to the huge area comprised by the Neotropics, one can imagine that forest conservation should not be an easy issue, requiring unmistakable strategies according to the distinctiveness on species composition and ecological traits of the vegetation. Indeed, heterogeneity of NF has been recognized for a long time, and many systems have been proposed for their classification. In a useful synthesis of such systems, five main forest physiognomies were proposed—broadleaved forest, mixed needle-broadleaved forest, stiff-leaved forest, broadleaved dwarf-forest, and stiff-leaved dwarf-forest—based on canopy structure and species

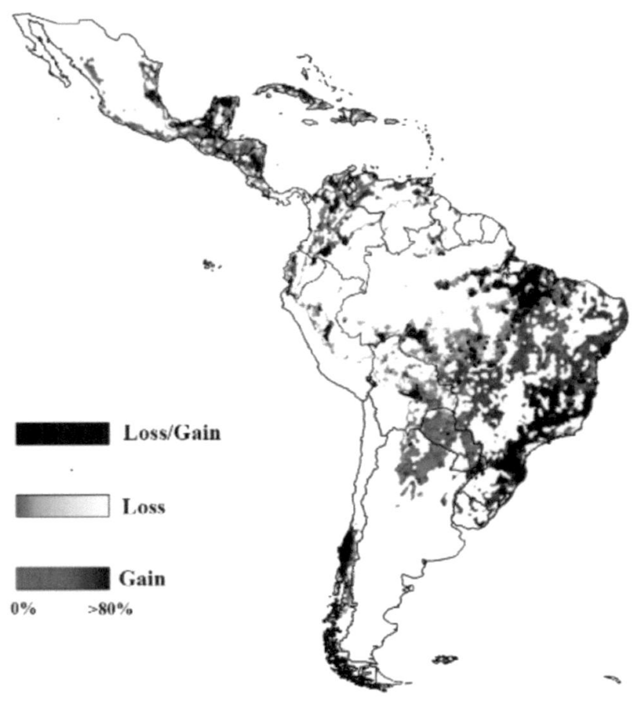

Figure 1. Extent of forest loss (red), gain (green), and both (magenta) for the 2000–2012 period, as stated by Hansen et al. (2013) [2].

assemblage [6]. The author, however, stresses that these criteria on their own are not enough to embrace all disparities of the vegetation and thus incorporates additional descriptors based on ecological and physiognomic attributes to be used in different combinations, as far as required. In consonance to this philosophy, many recent studies of biogeography, macroecology, evolution, and/or conservation recognize the average annual period for plant growth as the main attribute to delineate two main types of Tropical NFs, which we also adopt here: (a) seasonal dry forests (SDFs, are those occurring in regions with <1800 mm of annual rainfall and at least 5–6 months with less than 100 mm) [7] and (b) rain forests (RFs, are those occurring under more than 1800 m of mean annual precipitation, mean annual temperature >18°C, and seasonal variation in temperature lower than 7°C) [8]. It is important to highlight that water availability is determined not only by rainfall regime but also by variations of ground water regimes related to soils and geomorphology. Therefore, forests under different precipitation conditions may have similar growth periods, species compositions, and ecological traits [9]. It must be also stressed that the RF and SDF concepts are expanded here to include the southern temperate NF in spite of their strikingly distinct flora and environment [10].

In general, the larger continuous extents of NF are found as RF, although both patches and incursions are not uncommon. On the other hand, SDFs show a predominantly patchy distribution, with a wide variation in patch size [11]. Contrasting with SDFs, RFs display a lower species turnover, despite the intense plant migration favored by the instable conditions of those very dynamic communities. Notwithstanding the impressive α-diversity, the β-diversity of RF is low compared to that of SDF. Besides this, the γ-diversity in RF is high, due to the variation on species pools among geographically isolated patches. Opposed to this, the more stable SDF have a low rate of migration events, leading to higher levels of endemism. Despite the much lower α-diversity of SDF, the genetic variation of populations is commonly higher.

It is unfortunate that a low percentage of the Neotropical forests is under legal protection, with less than a quarter of the RF and just 4% of SDF, and, awkwardly, both of them are at present-day agriculture and livestock frontiers [12–14]. While the high turnover of species in continental, regional, and local scales is an inherent challenge for biodiversity conservation of the SDF, the growing fragmentation is a current threat to the maintenance of diversity and recovery of degraded areas in both NF and RF [15]. Additionally, current environmental changes—such as global warming associated to increased drought severity—may boost disturbance events and tree mortality and lead to the development of novel ecosystems, where hitherto NFs were dominant. Predicting how species and communities will behave in this scenario is thus a priority for successful actions of conservation of both Neotropical RF and SDF in the next decades.

2. Evolution of the Neotropical forests

It is widely accepted that most of the NF flora has evolved from a Jurassic Gondwanan pool, where cool-temperate vegetation, constituted principally by a conifer-tree/fern forest community, was found [16]. By late Cretaceous, however, the radiation of flowering plants changed this flora to that of more or less broadleaved and evergreen forests, with a differentiation between the northern and southern Gondwana land masses. The Sub-Antarctic fraction of the super-continent was quickly and almost completely dominated by angiosperms, while the Austral held down elements of the temperate vegetation, giving rise to the so-called Gondwana Paleovegetation. While the first represents the pristine source of diversity for the present South American tropical forests, the latter is considered the ancestral of the present forests of the humid and cool-temperate south. Indeed, many ancient lineages of the mixed paleo-flora, as Araucariaceae, Cunoniaceae, Lauraceae, Myrtaceae, Monimiaceae, Nothofagaceae, and Podocarpaceae, can still be found in southern South America despite the drastic changes that took place during the Paleocene [8]. During the Cenozoic Paleocene-Eocene Thermal Maximum, the Austral-derived flora started a migration toward lower latitudes, reaching the southern borders of the tropics. This process continued during the subsequent warming events until the Early Eocene Climatic Optimum, when both temperate and tropical forests expanded and experienced intense *in situ* speciation events. Tropical plant diversity then rose incredibly, with speciation surpassing extinction at around 30%. Despite the high speciation rates in the tropics, such processes also took place independently along the latitudinal gradient beyond its limits, leading to a specially diversified set of clades. The already recognizable Neotropical rainforests were also expanding, with a few lineages reaching the extratropical South America, and, at the same time, elements of the Northern Hemisphere were also reaching the tropics after the uplift of the Mesoamerican isthmus [8, 17].

The warming events of the Eocene intensified the expansion of the tropical flora, but it is very likely that forest communities as a whole could not invade or coexist with the temperate forests, as the southern flora at that time lacked most of the clades that were abundant in the tropics (e.g., Anacardiaceae, Annonaceae, Euphorbiaceae, Meliaceae, Moraceae, Sapotaceae, Violaceae, and Zingiberales) but were rich in families rare anywhere else (e.g., Akaniaceae, Atherospermataceae, Casuarinaceae, Cunoniaceae, Myrtaceae, and Proteaceae) [8]. With the rise of the Andean Chain in the Neogene, new marked changes took place in the South America flora. In the South, many lineages became restrict to the pacific side of the mountains, as well as extinction processes occurred (e.g., *Casuarina* and *Eucalyptus*), and another expansion northwards was initiated, this time crossing the tropics frontier and occupying the recent emerged mountainous areas [18]. Regarding Amazon, the area became much more dynamic leading to the emergence of much more heterogeneous traits, which drove the diversification of many clades [16, 19]. Further climate changes along Neogene and Pleistocene drove the consolidation of the RF and SDF nuclei, despite their areas waxed and waned following the fluctuations of both rainfall and temperature. It is believed that before the glacial maximum of the late Pleistocene, the tropical Neotropics was mostly covered by forest formations, and that SDF widely dominated the Dry Diagonal during the Pleistocene Glacial Maximum. They were probably not continuous at that or any other time, which is reflected on the current scarcity of widespread species throughout SDF islands [7]. Yet, the presence of many populations at the same SDF nuclei for many million years indicates that core areas have been preserved over time. Nonetheless, it is very likely that SDF had expanded into RF domains along both the Pleistocene and the late Holocene Last Glacial Maximum, favored by the cooler and dryness of such periods. After the Last Glacial Maximum, tropical SDF suffered many events of expansion and contraction, being the broader expansion reached on the mid-Holocene dry interval (~5000 years ago) [9].

http://dx.doi.org/10.5772/intechopen.72608

Rain forest areas, unlike those of SDF, used to diminish during the Quaternary glaciations, when they became, at higher or lower levels, fragmentated and confined into refugia islands [20, 21]. The matrix surrounding these patches suffered varied alterations, with part remaining forest (SDF) and part being covered by other vegetation types, particularly grasslands and savannas. Migrations across South America indeed occurred during such periods, including the establishment of the Andean *Alnus* and *Podocarpus* into the central Amazonian lowlands. Because of the reduced temperatures, precipitation, and atmospheric CO_2 of the Last Glacial Maxima, the Amazonian forests were less productive, had lower canopy structure, and were floristically and distinct than those of today. The basin area was predominantly covered by forests and was affected in different ways by the climatic changes. Vegetation at the western catchment area remained part of the Intertropical Convergence Zone and was less impacted, while the southern/eastern RFs were liable to drier and longer dry seasons and were mostly replaced by SDF or savannas [22]. Fire had an important role on species control during the subsequent expansion and contraction cycles especially at the southern/eastern regions RF [23]. Such processes probably influenced the present gradient of RF biodiversity within the basin. On the Late Holocene, RF reached their widest distribution, as a response to the increased precipitation caused by greater austral summer insolation. Fire outbreaks increased in the drier periods of the Holocene, resulting in the exclusion of a set of species from the most affected areas and thus taking part in the modulation of the present-day biogeography of Amazonian. In the following dry periods of the early-mid Holocene, such forests contracted again until reaching the current configuration—or that of the European arrival [19, 24]. Rain forests of the Atlantic domain were also impacted by the Pleistocene and Holocene climatic fluctuations, provoking the emergence of many C4 lineages during the drier periods, especially at the Last Glacial Maxima. After that, Atlantic RF experienced a big expansion and had the representativeness of gymnosperms diminished, being likely, then, that Atlantic Forests reached its modern floristic composition on the early Holocene [25, 26].

The Late Quaternary Extinctions of the South American megafauna (~50,000–10,000 years ago) might also have contributed to changes on the NF vegetation and flora [27]. For example, the dispersal processes of the zoochorous species were probably affected, particularly for those bearing large fleshy fruits, perhaps leading to the extinction of part of them. Because most of these should have been of climax species, their decline could have affected the dynamics of forests as a whole. In addition, litter amounts probably increased with diminished herbivory, and, consequently, enlarged fuel loads and fire frequency/intensity—a process significantly raised during the last 9000 years by the human arrival in the Neotropics [28]. Quaternary fires might have partially and temporarily suppressed or disturbed specific regions of RF, which are nowadays less diverse. Contrary to what one may think, this process probably has not favored SDF establishment, which is indicated by the absence of fire adaptations in their present-day flora [7]. Nonetheless, the anthropogenic impacts that took place afterward certainly surpassed by far those promoted by fire, with significant suppression of both SDF and RF registered since the development of the first American cultures. Besides this, such processes were greatly intensified after Spanish and Portuguese occupation, when RF exploration was also initiated [13, 14]. Changes in such vegetations are still occurring in an accelerated rate, and increasing our knowledge on the response of NF to these threats is urgent whether we are legitimately interested on their conservation.

3. Neotropical forests: a complex vegetation mosaic

Notwithstanding the fact that the plant cover has gone through striking variation along the geological eras, forests were always well represented in the Neotropics. In fact, Europeans found massive extents of forests in both Meso and South Americas, when they arrived in those continents (**Figure 2**). The distribution pattern of both RF and SDF of that period can still be noted, although much of the plant mass has been depleted or modified in structure and connectivity. In general terms, RFs are found in areas with humid climates, independently of pedologic traits or flooding regimes [9]. Areas with prominent dry periods, depending on the existence of enough water supply and/or suitable pedologic conditions, may harbor either RF or SDF, respectively. Contrastingly, drier areas with low ground water and fertility are usually covered by either savannas or grasslands. Therefore, RFs are still found on the vast Amazon domain and on the Atlantic, Pacific, and Caribbean windward coastlands—the latter three much more fragmented than original [14]. Additionally, RFs are also found as riverine seasonal forests across drylands, as well as temperate rain forests in the Southern Cone [10, 29]. The scattered distribution SDF concentrates across the so-called Pleistocene Arc that surrounds the Amazon basin before stretching into Mesoamerica and the Caribbean [11, 15, 30].

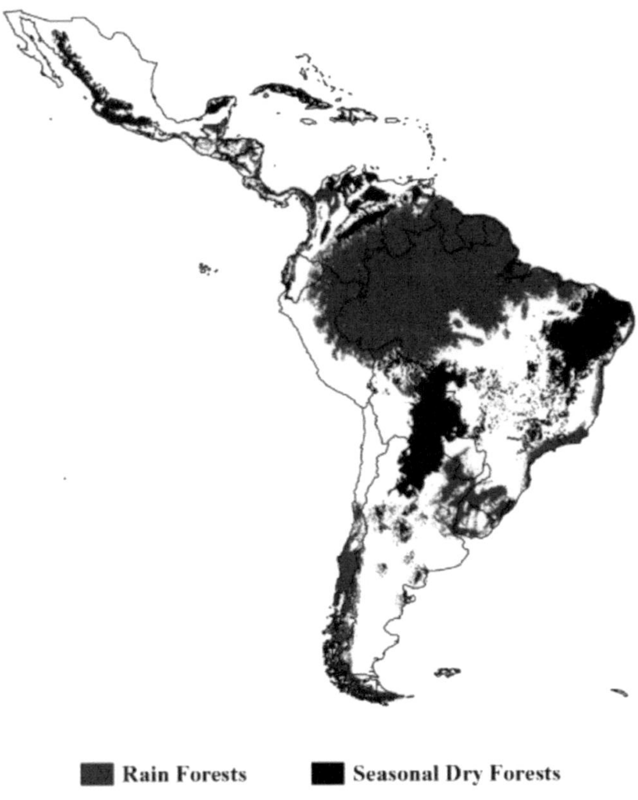

Figure 2. Extent and geographical distribution of Neotropical forest domains (solid patches; discontinuity within domains are not considered) and incursions in other domains (spotted areas). Adapted from Jaramillo and Cárdenas (2003), Fernandez-Vega et al.(2017) [14], and Rezende et al. (2017) [10].

Originally, RF covered more than half of Mesoamerican lands, where they had a somewhat discontinued distribution. Although human occupation increased their fragmentation, RF still occupies 30% of the isthmus area [14]. Regarding South America, almost half of the continent used to be covered by RF, with the Amazonian forest corresponding to a third of the continental area [8].

The Amazonian domain has a high α-diversity, though with a small set of trees (227 species) dominating all across [19]. Species composition and richness, however, vary widely in the region, with a notable increase in species richness toward the Andes [31]. Considering the botanical families, around 50% of the tree species belong to a small group formed by Leguminosae, Moraceae, Annonaceae, Euphorbiaceae, Lauraceae, Sapotaceae, Myristicaceae, and Arecaceae. Four out of these eight families are also well represented in terms of stem density, with around 60% of the trees belonging to Leguminosae, Arecaceae, Rubiaceae, Violaceae, Euphorbiaceae, Meliaceae, Sapotaceae, and Moraceae [8]. Some of these families are distributed in abundance gradients along the whole Amazonian region. For example, both Leguminosae and Burseraceae increase from the southwest to the northeast, where the well-represented Leguminosae may account for almost a quarter of the trees. In small scales, some families can show distinct distribution patterns, as that found for Lecythidaceae in French Guyana [32]. The second largest extent of RF is found along the South American Atlantic coast, namely the Atlantic domain, which extends from north-eastern Argentina and eastern Paraguay to north-eastern Brazil (**Figure 2**). In this domain, environmental traits are strongly influenced by the long latitudinal range, the mountainous terrain, as well as the sharp climatic gradient toward the continent hinterlands, conferring to the region more environmental contrasts than those across the whole Amazonian domain [33]. Those forests are home to a great β-diversity, resulting from high levels of both endemism and α-diversity, although a small number of species may be locally dominant [34]. Therefore, the Atlantic domain harbors not only RF but also other vegetation types, particularly where it finds its environmental limits. The following main types of RF may be distinguished accordingly: rain, cloud, rocky cloud, and Araucaria (mixed) forests, which are considered the core area of Atlantic forests. Semideciduous forests and rocky dwarf-forests are the predominant vegetations of the inner limits of the Atlantic domain, whose marginal habitats harbor almost half of the species endemic to the domain [29]. Many species from the Atlantic SDF and, less frequently, from the RF are found on the tropical and subtropical riverine forests that extend into the neighboring drier domains. Additionally, several species from the Amazonian RF are found in those riverine forests, which can be considered a floristic link among all disjunct South American RF [9, 35].

SDF are widely but discontinued distributed across Neotropics, and differently of RF, they lack both a latitudinal diversity richness gradient and many widespread species but always present a high β-diversity at both species and family levels. While they occupy almost a quarter of South America, they have a reduced representativeness in Mesoamerica, south of Mexico [7, 8]. Regarding the most representative families, Leguminosae hosts by far the largest number of species, followed by Burseraceae (particularly in Mexico), Cactaceae, Capparaceae, Erythroxylaceae, Malvaceae (Bombacoideae), and Zygophyllaceae (especially in Central America). Several clades have their species concentrated in or confined to Neotropical SDF. Many species are endemic to a specific region and belong to ancient lineages, most of which are monophyletic. Even when polyphyly is noted, the sister species are often found in the same region, indicating a high geographic phylogenetic structuration. Besides that, the endemics tend to be abundant, resulting in a metacommunity pattern of high mid abundance levels [7]. There is a notable discontinuity among SDF islands, apart from both extremes of the Dry Diagonal, namely the Caatinga and Chaco domains, in north-eastern and central-southern South America, respectively (**Figure 2**) [36]. Within SDF domains, soil traits can be very important to determinate the deciduousness degree, with evergreeness as an efficient strategy to save input nutrients under oligotrophic conditions [37].

It is believed that the Neotropical SDF of the South American Dry Diagonal had a much more continuous cover during the Pleistocene Glaciations [30]. Significant floristic differences are currently found among the three domains in the region, namely the Caatinga, Chaco, and Cerrado, which is the Brazilian savannas domain, where many incursions of SDF are found [36]. In some cases, the number of endemics can be high, as in Caatinga, within which a rocky portion can be discriminated due to the abundance of some species, belonging to four families: Cactaceae (*Brasilicereus phaeacanthus, C. albicaulis, Facheiroa squamosa, Pereskia aureiflora, P. bahiensis, P. stenantha, P. glaucochrous, P. pentaedrophorus, Stephanocereus leucostele*), Euphorbiaceae (*Cnidoscolus bahianus, C. argyrophylloides*, and *Jatropha palmatifolia*), Leguminosae (*Blanchetiodendron blanchetii, Mimosa irrigua, Poincianella laxiflora, Pseudopiptadenia brenanii, Pterocarpus villosus, P. zehntneri*, and *Pterodon abruptus*), and Rutaceae (*Galipea ciliata* and *Pilocarpus sulcatus*). In the case of the Cerrado, the number of generalists is increased because of the numerous tree species shared with the co-occurring savannas and riverine forests, which contribute to the higher species richness in the domain. A number of species are common throughout the Dry Diagonal, such as *Anadenanthera colubrina, Myracroduon urundeuva, Handroanthus impetiginosus, Aspidosperma pyrifolium, Senegalia polyphylla, Amburana cearensis, Schinus brasiliensis, Annona leptopetala*, and *Platymiscium floribundum*.

The remaining South American SDFs, occurring from northern Argentina to the Caribbean coast of Colombia and Venezuela, show similar patterns, despite the occurrence of a few distinct species groups. Venezuelan SDFs, for example, include *Bourreria cumanensis, Erythroxylon havanense, Guapira pacurero*, and several distinct species of Bignoniaceae, Boraginaceae, Capparidaceae, Euphorbiaceae, Leguminosae, Flacourtiaceae, Malpighiaceae, Rubiaceae, Rutaceae, and Sapindaceae [37]. However, there is also a great floristic variation as only a third of the botanical families are found throughout. This includes, for instance, increasing abundances toward the east of *Acacia tamarindifolia, Amaioua guianensis, Bourreria cumanensis, Bunchosia mollis, Bursera simaruba, Capparis verrucosa, C. tenuisiliqua, Cordia dentata, Croton choristopelis, Malpighia glabra, Piscidia carthaginensis, Pithecellobium oblongum*, and *Tabebuia chrysanta*, while *Acacia glomerosa, Amyris ignea, Chlorophora tinctoria, Fagara monophylla, Guaiacum officinale, Lonchocarpus violaceus, Morisonia americana, Trichilia hirta*, and *Vitex compressa* are very common in the north. The north also includes *Bourreria succulenta, Helietta pleeana, Krugiodendrum ferreum, Linociera caribaea, Pseudobombax septenatum, Zizyphus cinnamomum*, and the association of *Bursera* and *Lonchocarpus*, which are also very common in the Caribbean but rare across Venezuela.

RF and SDF floras of the temperate South America are strongly distinct of their correspondents in the tropics [8]. While Atherospermataceae, Winteraceae, Cunoniaceae, and Proteaceae are some of the most abundant families, they are scarce in the tropics. At the same time, many of the major tropical clades as Anacardiaceae, Annonaceae, Meliaceae, Moraceae, and Sapotaceae are absent in temperate lands. Monotypic genera are more well-represented in the southern flora, with several of them even belonging to monogeneric families such as Aextoxicaceae, Desfontainiaceae, Eucryphiaceae, and Gomortegaceae. In this way, many lineages, such as *Aextoxicon, Citronella, Cryptocarya, Drimys, Gomortega, Laurelia, Persea, Laureliopsis, Legrandia, Nothofagus, Pitavia,* and *Podocarpus,* have diversified outside the tropics, keeping their ancestral preference for temperate conditions. Although the southern floras are considerably poorer than their tropical counterparts, they have a higher lineage diversity, that is, they hold more diversity on broader groups, such as families, which means older diversification with higher conservatism of clades in the temperate floras [10, 18]. There is also a high conservation of ancestral genetic polymorphism in SDF, which is due to factors such as their arrested dynamics (see session V), low immigration rates, and notorious discontinued distribution, whose combination results in a long persistence of SDF populations and also increasing the endemism levels [11, 38]. Tropical RFs, otherwise, have been experimenting intense diversification in more recent times.

4. Neotropical forests floristics and diversity are controlled by environmental and anthropic traits

There is a high floristic uniqueness in both the tropical and temperate NF, with species richness by far higher in the former due to higher speciation and lower extinction rates overtime. On the other hand, the temperate flora has a higher lineage diversity, that is, richness of ancient clades [10]. Despite these contrasts between both tropical/temperate and RF/SDF floras, some patterns are common to all, especially with regard to family distributions. For example, warmer temperatures and higher rainfall increase species richness for both Leguminosae and Bignoniaceae, cooler temperatures for Asteraceae and Melastomataceae, and dryness for Polygonaceae, while abundance is favored by lower temperatures for Melastomataceae and Rubiaceae and higher precipitation for Arecaceae [39]. However, to our knowledge, there are no studies on the main modulators of the present-day clade distribution for the temperate Neotropics.

The main driver of forest type distribution across the Neotropics is the rainfall regime, although the temperature plays a major role in the subtropical and temperate sectors [18]. Indeed, rainfall itself controls forest structure, species richness, and successional dynamics. Together with soil fertility and landscape heterogeneity, temperature has also a positive correlation to taxonomic structural complexity, expressed as an increasing gradient of species, genera, and families richness from SDF to RF [35, 38, 40]. Species richness among SDF islands is poorly affected by changes in the amount of precipitation; otherwise, water restriction seems to be very important for the maintenance of SDF patches, preventing the establishment of RF [7, 35]. Surprisingly, despite the leading role of rainfall for most patterns in the tropical sector, variations in temperature are the main control for family diversity, in both abundance and species richness.

Several climatic features also modulate species distribution at and within regional levels. For example, temperature seasonality is the main controller of tree species composition in the subtropical sector of the Atlantic domain, particularly segregating Araucaria-dominated forests [18]. In the tropical sector, water deficit severity and mean annual precipitation are the best predictors of changes in species composition, segregating two main floristic groups, containing (a) rain, cloud, and cloud dwarf-forests and (b) riverine and semideciduous forests and campos rupestres [29]. Additional roles are played by fire and frost frequency in segregating the woody flora of rock outcrops, as does strong winds and salt-spray for coastal white-sand woodlands. Likewise, rainfall patterns are the main controllers of species richness across the Amazon domain, with higher figures in the much rainier west and center than in the seasonal and drier east and south [19, 31].

The existence of forests across the Dry Diagonal depends basically on two factors: higher soil moisture on valley bottoms harboring riverine seasonal forests and patches higher fertility soils covered on either deciduous or semideciduous forests [9]. On top of this, climate features play an additional role, affecting the species composition of SDF across the whole Diagonal, and extremes of cold temperatures and dry season severity are key factors. Annonaceae, Bignoniaceae, Leguminosae, and Moraceae, less able to cope with freezing temperatures, are more abundant and diverse in the warmer northern sector, while Capparaceae, Leguminosae, and Polygonaceae are favored in drier areas and Annonaceae, Arecaceae, Moraceae, and Rubiaceae in moister ones [30]. In addition, in the warmer northern sector, soil mineral nutrients are almost as important as ground water availability in determining changes in species composition [36].

Nonclimatic environmental traits, such as space, altitude, substrate, topography, soil composition, as well as anthropic impacts, are very important in determining NF types across other geographic regions. In this way, in north-western Argentina, the chief explanatory variables for species distribution on the Andean piedmont slopes are related to increasing moisture and decreasing temperature toward higher altitudes [41]. For example, the moister the area, the higher the abundance of *Diatenopteryx sorbifolia, Ocotea puberula, Cordia americana,* and *Eugenia uniflora,* while the opposite pattern is shown by *Calycophyllum multiflorum, Phyllostylon rhamnoides, Astronium urundeuva,* and *Anadenanthera colubrina.* In Venezuelan SDF, rainfall decrease is positively related to increasing abundance of Mimosoideae, Cactaceae, and Capparaceae (*Capparis hastata, C. linearis, C. odoratissima, C. flexuosa,* and *C. pachaca*), which harbor several physiological adaptations, such as relatively deep root systems and low hydric potentials [37].

At more local scales, where many environmental traits, such as rainfall regimes and soils, can be much homogeneous, the best predictors of beta-diversity are commonly land-form-related variables, as in the case of French Guyanese RF, where the distribution of plant taxa is mainly modulated by geomorphological features [32]. Lecythidaceae and Caesalpinioideae are predominant in the coastal plains and hilly hinterlands, while Burseraceae, Vochysiaceae, Simaroubaceae, and Mimosoideae are predominant in valley bottoms. Mimosoideae are more diverse and abundant in mountains, even surpassing Lauraceae, typically diverse in Amazonian mountains in general.

It is known that forest dynamics related to single and multiple-tree falls largely contribute to variations in tree species composition in RF. The emergence of gaps creates opportunities for the establishment of species belonging to different successional groups. Fast-growing and soft-wood tree species prevail immediately after gap creation and are followed by the emergence of the hard-wood species, which slowly become dominant in the canopy. However, these frequent and autogenic disturbances are not the only controlling factors of forest dynamics [14]. Indeed, other modes of disturbance may be important modulators of species composition, forest structure, and dynamics in certain sectors of the Neotropics. This includes severe ones, like hurricanes, convectional windstorms, severe droughts, fire outbreaks, floods, volcanic eruptions, landslides (caused by either rainstorms or seismic activities), and others of relatively moderate impact, such as droughts, fires, and shifting cultivation and regular winds. These impacts often interact, as in the case of wind frequency and strength, which increase in disturbed forests favored by discontinued canopy.

5. Modeling floristic and biomass dynamics in rain and dry forests

Forest communities along succession can be characterized according to the acquisitive or conservative traits of their populations, with regard to the economy of environmental resources. Conservative species can save limiting resources, such as water in drylands, while the acquisitive are not able to survive in restricted conditions of a given resource. In early successional stages, light is not a constraint factor in both RF and STDF, unlike water availability, which is a restrictive trait in SDF [38, 42]. Thus, conservative species are favored in the colonization of drylands, while there is no restriction to both functional types in humid areas. Anyway, acquisitive species are more competitive in RF because of their light-wood and related faster growth than that of the conservative hard-wood species. As succession goes on, RF environments get more light restrictive, favoring the growth of conservative shadow-tolerant species. On the other hand, for the less stratified SDF, warmness and dryness are both lessened after the pioneer community established, without significative restriction of light availability, which favors the arrival of the acquisitive species. As the conservative community grows up in RF, the light-demanding pioneer community starts declining, rather unlike the SDF where both functional groups are favored [38]. RF communities are very dynamic and, from the beginning of the death of conservative individuals, which starts earlier than in STDF, the proportion of functional groups is sustained overtime, despite their turnover on frequently opened areas. SDFs are more stable because of the increased longevity of their conservative individuals, and their gaps are smaller because of the smaller number of trees ripped off in the process. In addition, the re-sprout of neighboring living individuals is usual, while seedling survival rates are inexpressive. This scenario brings two main consequences for tree dispersal and species composition: (a) despite of the long-dispersal capacity of many SDF species, both dispersal of endemic species and the arrival of immigrants are hampered within their areas and (b) RFs are liable to both occurrences, and most of their species are not able to colonize SDF systems.

Such changes in functional composition and community structure are highly correlated to the above-ground biomass (AGB) dynamics, and as they are feedback-regulated by environmental traits, AGB dynamics is also distinct in RF and SDF. On early RF succession, when neither light nor humidity are limiting factors, the fast-growing light-wood species correspond to the majority of the AGB. Thus, the shadier the environment becomes, the more competitive become the hard-wood and slow-growing species, resulting on both loss of light-wood trees and accelerated rates of biomass accumulation. Before the proportion of functional groups is stabilized, increasing accumulation rates of AGB are still observed overtime, due to the growth of hard-wood trees. On the other hand, after the establishment of the pioneer and long-lasting hard-wood SDF community, successional events give rise to either the arrival of small amounts of light-wood trees or the sprouting of the slow-growing ones [11, 38]. Thus, there are two factors involved on differentiation of biomass dynamics: (a) both earlier and later succession groups in each forest type belong to different functional groups; and (b) while the pioneer community is quickly replaced in RF, mortality rates of hard-wood early community in SDF are low until most advanced regeneration phases. The combination of these two factors create divergent patterns for both individual biomass contribution and the increase on AGB rates overtime: (a) RFs present continuously rising gains in AGB accumulation during a long time, until very late-successional phases, when the mortality of some hard-wood individuals reduces those gains and (b) the rhythm of biomass accumulation in SDF is slowly accelerated since the early successional phases, and this acceleration becomes irregular in the more advanced phases, when increases in AGB accumulation rates due to both colonization and sprout may not surpass the decrease due to loss of old hard-wood trees.

Since the abundance of functional groups is strongly dependent on tree recruitment, growth, and mortality, these are also important drivers for biomass dynamics. AGB gains due to tree growth is the key factor on the modulation of biomass increase in intact RF and SDF, corresponding to almost 50% of the variation within all successional groups [38]. The biomass loss due to mortality has a secondary role, and, as described above, it is important since early successional phases of RF, as well as in late regeneration phases of SDF. Notwithstanding, since these processes are under environmental control, their importance in community dynamics is currently being altered by the recent climatic shifts. Brienen et al. [43] claimed that the elevation of atmospheric C concentration overtime is inducing a continued gain in productivity in Amazonian forests and accelerating the individual life cycles, leading to a great rise in the number of dying trees. Indeed, the authors reported that this process is the major contributor to the decrease in the C sink capacity of such vegetation since the 1990s. Nonetheless, despite of the fact that the average basal area of that vegetation decreased overtime, the importance of big trees death for reducing AGB accumulation rates is not so high. On the other hand, they also outlined that the exceptional dryness of both 2005 and 2010 should be responsible for the increase of the mortality of large-stature trees in Amazonia in those years. In the same way, [44] demonstrated that annual seasonality is directly linked to fluctuations of recruitment rates in Mexican SDF, which were higher during the moist season, when gains on biodiversity were also noted. However, increases in recruitment, growth, and species richness were not observed in those SDF during the rainy 2006 and 2007 years that followed the dry 2005. It is likely, then, that severe droughts lead to the depletion of water and carbohydrates and that some time is required for full recovery. It is probable too that such climatic events may lead to shifts in ecological relations as herbivory, pollination, and dispersal that also modulate recruitment, growth, and species composition of plant communities.

6. Environmental and anthropic drivers of biomass dynamics in Neotropical forests

Taxonomic diversity is a strong modulator of AGB pools along tropical forests. Together with the functional dominance, which is the ponderation between the average of a functional trait and the relative abundance of a given clade in the community, taxonomic diversity explains almost 40% of the differences in the tropical forests C pools [45]. Along the Neotropics, however, besides the great variety in biomass and biodiversity, the correlation between them is weak. In this way, the species composition along NF, as inferred from its correlation with wood density functional dominance, is not correlated to AGB stocks within the vegetation. Anyway, this correlation can be observed at small scales because of niche complementarity in areas smaller than 1 ha [31]. Nevertheless, community structure, represented by the maximum diameter functional dominance, is a key driver for the differentiation of RF and SDF according to their C storage potentials [38]. Likewise, there is a high variation of forest structure in regional scales within these forest types, and, in combination with other traits, it can also control their AGB. An interesting example is found for Amazonian intact forests: (a) the more dynamic tree communities are, the lower is their average wood density; (b) the less dynamic peripheral areas also show lower average basal areas; (c) the integration of these features produces a gradient of decreasing AGB toward those peripheries [46]. Curiously, taxonomic diversity richness distribution follows this same biomass gradient [19], but further investigation is needed to confirm this as a pattern for the Neotropical RF.

NF cover is currently undergoing unprecedented changes. For instance, while forest regeneration is observed in many sites of highly impacted RF (**Figure 1**), appalling deforestation rates grants SDF the title of "most supressed woody ecosystem in the world" [2]. Anyway, together with the increasing number of degraded areas, the sites in regeneration are also increasing, boosted by their high regrowth potential [47]. As an example, C sequestration rates by secondary NF in 8 years were more than 10 times higher than that of the preserved Amazonian RF, reaching impressive rates of 3.05 Mg C ha^{-1} $year^{-1}$ in average. Likewise, in a period of two decades, secondary forests were responsible for the rise of 122 Mg C ha^{-1} on the total AGB in the Neotropics, then becoming the major global contributor as a C global sink, despite the usually higher AGB pools of mature forests [48, 49]. In this way, a scenario of increased reforestation would easily improve the Neotropical C sink capacity. However, the actual situation is not so auspicious. For example, during the century, while C sequestration by secondary forests rose on a rate of 0.05 Pg C $year^{-1}$, the contribution of well-established forests to the C sink was reduced in 0.23 Pg C $year^{-1}$, and C emissions by deforestation increased on a rate of 0.14 Pg C $year^{-1}$ [3].

Land use history has a key role on NF recovery rates. In this way, areas with denser tree residual tend to show quicker forest resurgences than those with no arboreal cover [12]. At the same time, the potential to increase the biomass pools also depends on the percent of forest cover of the surrounding matrix, the proximity to other fragments, and the time of regeneration [50]. In addition to this, environmental traits as rainfall regime, and other climatic features related to it, are key modulators of biomass, because they directly influence the growing season, while soil traits play a supporting role, modulating the possibilities of returning to the climax structure [48, 51]. Much varied rates of AGB recovery are found among NF, both between RF and SDF, and within them. For example, rates from 4.7 to 6.1 Mg ha^{-1} $year^{-1}$ in RF and from 2.3 to 1.9 Mg $year^{-1}$ in SDF were observed, with the average AGB variation corresponding to shorter and longer periods, respectively [38]. Interestingly, functional traits found for SDF communities were claimed as positively correlated to regeneration of NF as a whole [48]. For the authors, the abundance of old tress belonging to hard-wood pioneer species in disturbed sites would be positive modulators of the time and rates of biomass recovering, while data of a Mexican SDF, where no clear correlation was found between regrowth span and regeneration rates, also suggest that it is not likely that time is an important controller of this vegetation [44]. Otherwise, the community structure at the beginning of the recovering process, as expressed by initial AGB or basal area, was found as strong modulators of biomass increases in RF, also influencing both the litter productivity and decomposition [31, 52].

7. Neotropical forests facing the Anthropocene era

Great human impacts on NF are recorded since the beginning of the expansion of the first American civilizations, especially in Mesoamerica, with an expressive vegetation suppression and fragmentation both of which were much intensified after the arrival of the Europeans at the Neotropics [14]. Nonetheless, both deforestation and fragmentation rates were never as high as they are in this millennium (**Figure 1**), particularly following the advance of the agricultural frontiers [3]. This is a worrying scenario whether C sequestration is considered, since a third of the whole global C sink capacity is attributed to well-preserved NF [45]. In addition, this potential has been decreasing, as stated for Amazonian intact forests, which showed the C sink capacity reduced in a third during the millennia transition [43]. The decrease is even sharper in sites where standing forests are managed, for example for timber extraction, both directly because of massive withdraw of hard-wood tress and indirectly because the process leads to loss of seed matrices [3]. For instance, average biomass accumulation rates in logged areas of Amazonia were lower than a half in comparison to areas under natural regeneration in the same region [48]. Fortunately, areas of fast-growing secondary forests are becoming widespread, both across established forests and at their neighborhoods, with a trend of becoming more important in the upcoming sceneries [53]. Then, from the point of view of integrity, NFs are currently a dynamic mosaic of successional phases, where continuous areas with intact vegetation are decreasing, while both fragmentation and natural or controlled regeneration are increasing (**Figure 1**).

Since forest resilience can be expressed in terms of floristic, functional, and biomass dynamics, isolation of areas leads to severe impacts, such as biodiversity loss and reduced C sink capacity [8]. Deforestation and/or fragmentation promote the largely studied edge effects, whose impacts affect the forest interior up to tens to thousands of meters, as does the so-called "vegetation breeze," widely registered for Amazonian forests [54]. This phenomenon is a result of the difference between microclimates of forests and nonforest areas, which leads forest environments to varied levels of desiccation. In this way, the drier and warmer the adjacent area, the more impactful is the disturbance, with areas under regeneration, then, playing an important role in mitigating the process. Additional edge effects are noted when forest suppression is combined with fire occurrences [55]. In these cases, atmospheric water droplets may be condensed to aerosol particles, which hamper the formation of proper water drops, intensifying the dryness in the region. Perhaps the most immediate outcome of this whole process is biodiversity loss, with even the most diverse fragments showing significant decrease of abundance/diversity of both fauna and nonarboreal plants, such as lianas and epiphytes [34]. Likewise, only large fragments may harbor plants that require a large area or specific habitats, which, however, does not mean that small forest patches have no ecological importance [33, 54]. In fact, isolation of fragments hampers biodiversity conservation and functional dynamics maintenance, and, because of this, all fragments may have an important role both as eventual reservoirs for plant species and as stepping stones for fauna, processes which are more effective at shorter distances from well-conserved forest remains [34, 53].

Climatic traits are also important modulators of forest resilience (see sessions IV and VI), since they strongly control physiologic traits, which, in their turn, are responsible for controlling individual processes, with important ecological implications, such as tree growth and mortality [8]. Physiologic responses leading to mortality include both increases on photorespiration oxidative stress and C starvation and hydraulic failure, all of them directly or indirectly linked to reduced differences between soil and leaves hydric potentials [56]. Despite the experimental difficulty for studies on the theme, many inferences have been done, especially for the better-studied Amazonian forests. For instance, biophysical processes, such as mechanical and hydraulic failure, have been claimed as the main inducers of large-trees mortality in Amazonia during the last 30 years, together with their accelerated growth in response to CO_2-fertilization [43]. Such increased growth would be linked to the ability of plants under higher concentration of atmospheric C to maintain their stomata closed during a longer period and, in this way, avoid desiccation [54]. Nevertheless, this mitigation mechanism might not avoid increases on embolism occurrences, since acquisitive water use strategies of RF usually include cheap transport throughout larger vases and leaves lacking very negative hydric potential [56]. On the other hand, ecophysiologic responses may also mitigate oxidative stress, as it was claimed for Amazonian forests, where increased emissions of biogenic volatile organic compounds, such as isoprene, in response to warmer temperatures, should have led to reduced tree mortality rates [49]. Regarding drought and DF species, the ability of controlling C absorption would favor functional groups in different ways. Deciduous species would be more competitive under intensified and short dryness, while the evergreens would show lower mortality rates under prolonged but weak droughts because they can control stomatal opening and avoid C starvation [51].

Intensified dryness as an effect of less and/or more variable rainfall are expected to be the worst outcome of the predicted climatic changes for both biomass and community dynamics in NF, since increases in temperature were not correlated to biodiversity losses during the former global warming [19]. However, warming and more severe droughts are already in course, with field observations, most of them for RF (see sessions IV and VI) confirming that forest resilience is indeed correlated with water availability. In spite of the lack of information for other forests, experimental data for the Amazonian show that their resilience can be high, since most of species can stand until 60% of soil hydric content losses and significantly maintain their survival rates [19]. It is not known, for example, whether SDF species will survive under new climatic conditions at their own current geographic location, since these can become very harsh and perhaps incurring in the desertification of those regions [57]. Likewise, some authors believe that changes in climatic traits in some RF regions, such as eastern Amazonia, will both disturb the stability of the local community and favor the arrival of SDF/savanna species, depending on soil fertility and fire frequency and intensity [51], while others claim that warmer temperatures will lead to increased mortality at high altitudes and boost the relative abundances of lowland tropical species [49]. Anyway, dispersal is a potential limiting for species migration across the Neotropics, since both defaunation and fragmentation are in course processes under accelerated rates.

Notwithstanding, continued alterations in both species distribution and functional composition are expected for NF up to the end of this century. Such alterations are likely to result in reduced carbon storage, as observed for Amazonia, while some authors believe that resilience of natural or man-induced secondary regeneration will decrease (see session VI) [51]. One reason, with an optimistic hope, from the point of view of C sequestration, perhaps that the vegetation cover of regions currently harboring RF will change to SDF, while the opposite will not happen. As SDF species are slow growing, which may be even accentuated under drier conditions, rates of C sink would decrease in all regions. The application of climatic projections to predict future species range is usual, but it is not an easy work predicting changes in community composition and forest structure, from which C dynamics is calculated [58]. Then, computer simulations for NF biomass and floristic dynamics show divergent results, sometimes confirming the above-cited scenarios. For example, tropical NF will have the floristics highly affected during the next decades, with the mortality of big trees responding for almost 30% of AGB loss, while lowland Amazonian forests will be partially replaced probably by SDF or, less likely, by savannas [23]. Meanwhile, secondary lowland NF will provide a large C sequestration for the next 40 years, equivalent to all the C emissions by the whole Neotropics during the last two decades [47] (**Figure 3**), perhaps favored by CO_2-fertilization, which combined with the reduction of the current levels of deforestation, would lead Amazonian forests' structure to remain almost intact even if atmospheric CO_2 levels are double the present [54]. The larger consensus among researchers, however, is that the more information such models incorporate as independent variables, the more trustworthy will be the simulated NF responses to climate changes [49, 58]. In this way, models should incorporate answers for some basic questions [49, 53]: (a) which will be the forest types more affect by dryness and how will they be affected? (b) how will fauna (dispersers, predators, etc) distribution be affected? (c) which plant physiologic traits will enhance or mitigate both dispersal and mortality under the new scenarios? (d) why are both large and small trees affected in different ways by drought? (e) how different species are affected by the all these processes?, and (f) in which ways does secondary forests growth affect the neighbor environmental conditions?

Some of those questions, or at least part of them, are already being answered, for example, how shifts in tree community composition lead to alterations of both fungal diversity and the linked vegetation drought resistance [23]. Others, such as arboreal community vigor during seasonal or eventual dryness, may be correlated to remote-sense observational data, which could easily be incorporated as independent variables in models [49]. Additional predictors to be incorporated are described in the sessions IV and VI of this chapter, including how

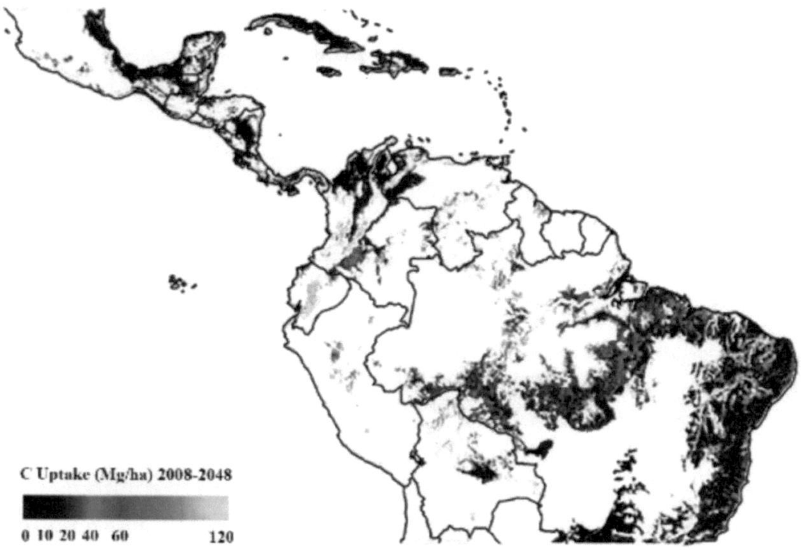

Figure 3. Estimated C sequestration by lowland Neotropical secondary forests (<1000 m), as stated by Chazdon et al. (2016) [47].

fragmentation modulates the neighboring forest environments as a response of vegetation breeze, as well as which is the correlation of logging intensity and biomass recovering overtime, and how floristics dynamics are affected by drought. Most of these studies, however, were settled in Amazonian forests, remaining a huge lack of information for the other RF and all SDF [51], with the exception of the environmental traits modulating floristics, which are available for most NF (session IV). In this way, our knowledge regarding the future changes on NF carbon storage leading to a carbon-cycle feedback is still largely uncertain [58]. Thus, if we want to improve such predictions, urgent goals need to be set, such as expanding the existing information for the all NF types and also investigating other theories, as those exemplified in the first paragraph of this session, which requires massive investments in long-term observations and experiments on the ground across the whole Neotropics.

8. Conclusions

The potential major role of NF for mitigating the increased levels of anthropogenic atmospheric carbon dioxide is undebatable. Despite the few certainties about their ecological responses facing global climatic changes, it is vital to use the current knowledge to both conserving and recovering those forests. Survival of such invaluable forests, from the point of view of biodiversity maintenance and/or C sink, is directly dependent on decreasing deforestation rates. Forest suppression is already inducing expressive impacts such as the increase of 1 month on dry season at the arc-of-deforestation, already observed, and the expected severe disturbs on the neighboring south-eastern Amazonia, both on vegetation and hydrology [55]. Forests historically more disturbed may even disappear under continued deforestation, as claimed for the much endangered SDF in both the Caatinga and Cerrado domains [36, 59]. Ironically, if SDFs were widely distributed across the Dry Diagonal during the drier periods of the Pleistocene, it is likely that water-conservative species of these domains should also survive under the supposed dryer future conditions in their neighboring regions and work as biological sources for their colonization. Considering a scenario of no anthropic intervention for decreasing SDF high fragmentation, migration across lands would be very unlikely to occur, while surviving riverine SDF would adopt a major role in such dispersal-dependent processes. Likewise, a second but not less important role for this corridor vegetation is that of a species reservoir, which would be also the case of the western Amazonia, and all other RFs under maritime moisture influence.

Improving both cover extent and biodiversity/biomass is also of central importance for the maintenance of forest environmental services. Some regions deserve special attention, as exemplified by both the ecotonal areas with drier domains, and wide and well-conserved old forests borders. In such places, reforestation should be massive, while within the domains, at least the number of stepping stones should be increased. Active intervention may be required for secondary forests according to their conservation goals [53], such as the introduction of late-successional species in isolated fragments with the objective of C sequestration and/or biodiversity conservation. Likewise, reduced moist areas would be the most appropriate sites for the establishment of managed-logging stands, which should be done under serious criteria, and preferentially surrounding existing forests, aiming at the conservation of their environmental traits.

9. Final considerations

In a general manner, it is very unlikely that NF will conserve their historical ecosystems and hold the prevalent past environmental and biota proprieties [60]. Instead, the replacement of those forests for either hybrid or novel ecosystems is more likely. However, environmental services such as nutrient, including C, cycling rates, do not necessarily have to change, but if we want to avoid this, prompt habit changes are need. Poor logging practices, forest suppression, and extensive monocultures are just few examples of the unhealth human relationship with nature. On the other hand, forest management practices already in course in many sites must be expanded to the whole Neotropics. Some of those recovering forests not only harbor a great biodiversity, for example those in the Atlantic domain [53], but also sink considerable C amounts (**Figure 3**). Anyway, the major forest regrowth polices worldwide always included both economic incentives and law enforcement, and the improvement of both practices is welcomed in the Neotropics, whether an environmental collapse is not desired

Author details

Cristiane Dias e Sarmento* and Marcel Giovanni Costa Franca

*Address all correspondence to: cristianedsarmento@hotmail.com

Instituto de Ciências Biológicas, Universidade Federal de Minas Gerais, Belo Horizontes, Minas Gerais, Brazil

References

[1] Crowther TW, Glick HB, Coveu KR, et al. Mapping tree density at a global scale. Nature. 2015;**525**(7568):201-205. DOI: 10.1038/nature14967

[2] Hansen MCC, Potapov PV, Moore R, et al. A large and persistent carbon sink in the World's forests. Science. 2013;**342**(November):850-854. DOI: 10.1126/science.1244693

[3] Pan Y, Birdsey RA, Fang J. A Large and Persistent carbon sink in the World's forests. Science. 2011;**333**(6045):948-993. DOI: 10.1126/science.1201609

[4] Saatchi SS, Harris NL, Brown S, et al. Benchmark map of forest carbon stocks in tropical. International Water Power and Dam Construction. 2011;**65**(8):30-32. DOI: 10.1073/pnas.1019576108

[5] Myers N, Mittermeier RA, Mittermeier CG, et al. Biodiversity hotspots for conservation priorities. Nature. 2000;**403**(February):853-858

[6] Oliveira-Filho AT. Um sistema de classificação fisionômico-ecológico da vegetação neotropical: segunda aproximação. In: Eisenlhor PV, editors. Fitossociologia no Brasil: Métodos e estudeos de caso. 2nd ed. Vicosa: Editora UFV; 2015. pp. 452-474

[7] Pennington RT, Lavin M, Oliveira-Filho AT. Woody Plant diversity, evolution, and ecology in the tropics: Perspectives from seasonally dry tropical forests. Annual Review of Ecology, Evolution, and Systematics. 2009;**40**(1):437-457. DOI: 10.1146/annurev.ecolsys.110308.120327

[8] Jaramillo C, Cárdenas A. Global warming and neotropical rainforests: A historical perspective. Annual Review of Earth and Planetary Sciences. 2013;**41**(1):741-766. DOI: 0.1146/annurev-earth-042711-105403

[9] Oliveira-Filho AT, Ratter J. Global warming and Neotropical rainforests: A historical perspective. Edinburgh Journal of Botany. 1995;**52**(2):141. DOI: 10.1017/S0960428600000949

[10] Rezende VL, Dexter KG, Pennington RT, et al. Geographical variation in the evolutionary diversity of tree communities across southern South America. Journal of Biogeography. 2017;**44**(10):2365-2375. DOI: 10.1111/jbi.13013

[11] Pennington RT, Lavin M. The contrasting nature of woody plant species in different neotropical forest biomes reflects differences in ecological stability. New Phytologist. 2016;**210**(1):25-37. DOI: 10.1111/nph.13724

[12] Rutishauser E, Hérault B, Baraloto C, et al. Rapid tree carbon stock recovery in managed Amazonian forests. Current Biology. 2015;**25**(15):R787-R788. DOI: 10.1016/j.cub.2015.07.034

[13] Calvo-Rodriguez S, Sanchez-Azofeifa AG, Duran SM, et al. Assessing ecosystem services in Neotropical dry forests: A systematic review. Environmental Conservation. 2017;**44**(01):34-43. DOI: 10.1017/S0376892916000400

[14] Fernadez-Vega J, Covey KR, Ashton MS. Tamm review: Large-scale infrequent disturbances and their role in regenerating shade-intolerant tree species in Mesoamerican

rainforests: Implications for sustainable forest management. Forest Ecology and Management. 2017;**395**:48-68. DOI: 10.1016/j.foreco.2017.03.025

[15] Banda KR, Delgado-Salinas A, Dexter KG, et al. Plant diversity patterns in neotropical dry forests and their conservation implications. Science. 2016;**353**:1383-1387

[16] Segovia RA, Arnesto JJ. The Gondwanan legacy in South American biogeography. Journal of Biogeography. 2015;**42**(2):209-217. DOI: 10.1111/jbi.12459

[17] Xing Y, Gandolfo MA, Jarenkow JA. The Cenozoic biogeographical evolution of woody angiosperms inferred from fossil distributions. Global Ecology and Biogeography. 2015;**24**(11):1290-1301. DOI: 10.1111/geb.12383

[18] Oliveira-Filho AT, Budke JC, Jarenkow JA. Delving into the variations in tree species composition and richness across South American subtropical Atlantic and Pampean forests. Journal of Plant Ecology. 2015;**8**(3):242-260. DOI: 10.1093/jpe/rtt058

[19] Olivares I, Svenning JC, Bodegom PM, et al. Effects of warming and drought on the vegetation and plant diversity in the Amazon Basin. Botanical Review. 2015;**81**(1):42-69. DOI: 10.1007/s12229-014-9149-8

[20] Mayle FE. Millennial-scale dynamics of Southern Amazonian rain forests. Science. 2000;**290**(5500):2291-2294. DOI: 10.1126/science.290.5500.2291

[21] Mayle FE, Beerling DJ, Gosling WD. Responses of Amazonian ecosystems to climatic and atmospheric carbon dioxide changes since the last glacial maximum. Philosophical Transactions of the Royal Society B: Biological Sciences. 2004;**359**(1443):499-514. DOI: 10.1098/rstb.2003.1434

[22] Anhuf D, Ledru MP, Behling H, et al. Paleo-environmental change in Amazonian and African rainforest during the LGM. Palaeogeography, Palaeoclimatology, Palaeoecology. 2006;**239**(3-4):510-527. DOI: 10.1016/j.palaeo.2006.01.017

[23] Bonal D, Burban B, Stahl C, et al. The response of tropical rainforests to drought—Lessons from recent research and future prospects. Annals of Forest Science. 2015;**73**(1):27-44. DOI: 10.1007/s13595-015-0522-5

[24] Malahado ACM, Pires GF, Costa MH. Cerrado conservation is essential to protect the Amazon rainforest. BioOne. 2010;**39**(8):580-584. DOI: 10.1007/s13280-010-0084-6

[25] Saia SE, Pessenda LCR, Gouveia SEM, et al. Last glacial maximum (LGM) vegetation changes in the Atlantic Forest, southeastern Brazil. Quaternary International. 2008;**184**(1):195-201. DOI: 10.1016/j.quaint.2007.06.029

[26] Pessenda LCR, Oliveira PE, Aravena R, et al. The evolution of a tropical rainforest/grassland mosaic in Southeastern Brazil since 28,000 14C yr BP based on carbon isotopes and pollen records. Quaternary Research. 2009;**71**(3):347-352. DOI: 10.1016/j.yqres.2009.01.008

[27] Gill JL. Quaternary research. New Phytologist. 2014;**201**(4):1163-1169. DOI: 10.1111/nph.12576

[28] Pessenda LCR, Gouveia SEM, Aravena R, et al. Holocene fire and vegetation changes in southeastern Brazil as deduced from fossil charcoal and soil carbon isotopes. Quaternary International. 2004;**114**(1):35-43. DOI: 10.1016/S1040-6182(03)00040-5

[29] Neves DM, Dexter KG, Pennington RT, et al. Dissecting a biodiversity hotspot: The importance of environmentally marginal habitats in the Atlantic Forest domain of South America. Diversity and Distributions. 2017;**23**(8):898-909. DOI: 10.1111/ddi.12581

[30] Neves DM, Dexter KG, Pennington RT, et al. Environmental and historical controls of floristic composition across the South American dry diagonal. Journal of Biogeography. 2015;**42**(8):1566-1576. DOI: 10.1111/jbi.12529

[31] Sullivan MJP, Talbot J, Lewis SL, et al. Diversity and carbon storage across the tropical forest biome. Scientific Reports. 2017;**7**(39102):1-12. DOI: 10.1038/srep39102

[32] Guitet S, Pélissier R, Brunaux OETAL. Geomorphological landscape features explain floristic patterns in French Guiana rainforest. Biodiversity and Conservation. 2015;**24**(5): 1215-1237. DOI: 10.1007/s10531-014-0854-8

[33] Ribeiro MC, Metzger JP, Martensen AC, et al. The Brazilian Atlantic Forest: How much is left, and how is the remaining forest distributed? Implications for conservation. Biological Conservation. 2009;**142**(6):1141-1156. DOI: 10.1016/j.biocon.2009.02.021

[34] Maçaneiro JPO, Oliveira LZ, Eisnelohr PV, et al. Paradox between species diversity and conservation. Tropical Conservation Science. 2016;**9**(4):1-19. DOI: 10.1177/1940082916668011

[35] Gentry AH. Diversity and floristic composition of neotropical dry forests. In: BullocK HS, Mooney HA, editors. Seasonally Dry Tropical Forests. 1st ed. Cambridge: Cambridge University Press; 1995. pp. 146-194. DOI: 10.1017/CBO9780511753398.007

[36] Santos RM, Oliveira-Filho AT, Eisenlohr PV, et al. Identity and relationships of the Arboreal Caatinga among other floristic units of seasonally dry tropical forests (SDTFs) of north-eastern and Central Brazil. Ecology and Evolution. 2012;**2**(2):409-428. DOI: 10.1002/ece3.91

[37] Fajardo L, Gonzales V, Nassar JM, et al. Tropical dry forests of Venezuela: Characterisation and current conservation status. Biotropica. 2005;**37**(4):531-546. DOI: 10.1111/j. 1744-7429.2005.00071.x

[38] Rozendaal DMA, Chazdon RL, Arreola-Villa F, et al. Demographic drivers of aboveground biomass dynamics during secondary succession in Neotropical dry and wet forests. Ecosystems. 2017;**20**(2):340-353. DOI: 10.1007/s10021-016-0029-4

[39] Punyasena SW, Eshel G, McElwain JC. The influence of climate on the spatial patterning of Neotropical plant families. Journal of Biogeography. 2008;**35**(1):117-130. DOI: 10.1111/j.1365-2699.2007.01773.x

[40] Bueno ML, Rezende VL, Pontara V, et al. Floristic distributional patterns in a diverse ecotonal area in South America. Plant Ecology. 2017;**218**(10):1171-1186. DOI: 10.1007/s11258-017-0759-1

[41] Blundo C, Malizia LR, Blake JC, et al. Tree species distribution in Andean forests: Influence of regional and local factors. Journal of Tropical Ecology. 2012;**28**(1):83-95. DOI: 10.1017/S0266467411000617

[42] Lohbeck M, Poorter L, Lebrija-Trejos E, et al. Successional changes in functional composition contrast for dry and wet tropical forest. Ecology. 2013;**94**(6):211-1216. DOI: 10.1890/12-1850.1

[43] Brienen RJW, Phillips OL, Feldpausch TR. Long-term decline of the Amazon carbon sink. Nature. 215;**519**(7543):344-348. DOI: 10.1038/nature14283

[44] Maza-Villalobos S, Poorter L, Martínez-Ramos M. Effects of ENSO and temporal rainfall variation on the dynamics of successional communities in old-field succession of a tropical dry forest. PLoS One. 2013;**8**(12):1-12. DOI: 10.1371/journal.pone.0082040

[45] Cavanaugh KC, Gosnell JS, Davis SL, et al. Carbon storage in tropical forests correlates with taxonomic diversity and functional dominance on a global scale. Global Ecology and Biogeography. 214;**23**(5):563-573. DOI: 10.1111/geb.12143

[46] Malhi Y, Wood D, Baker TR, et al. The regional variation of aboveground live biomass in old-growth Amazonian forests. Global Change Biology. 2006;**12**(7):1107-1138. DOI: 10.1111/j.1365-2486.2006.01120.x

[47] Chazdon RL, Broadbent EN, Rozendaal DMA, et al. Carbon sequestration potential of second-growth forest regeneration in the Latin American tropics. Advanced Sciences. 2016;**2**(5):e1501639. DOI: 10.1126/sciadv.1501639

[48] Poorter L, Bongers F, Aide TM, et al. Biomass resilience of Neotropical secondary forests. Nature. 2016;**530**(7589):211-214. DOI: 10.1038/nature16512

[49] Holm JA, Kueppers LM, Chambers JQ. Novel tropical forests: Response to global change. New Phytologist. 2017;**213**(3):988-992. DOI: 10.1111/nph.14407

[50] Ferraz SFB, Ferraz KMPMB, Cassiano CC, et al. How good are tropical forest patches for ecosystem services provisioning? Landscape Ecology. 2014;**29**(2):187-200. DOI: 10.1007/s10980-014-9988-z

[51] Allen K, Dupuy JM, Gei MG, et al. Will seasonally dry tropical forests be sensitive or resistant to future changes in rainfall regimes? Environmental Research Letters. 2017;**12**(2):023001. DOI: 10.1088/1748-9326/aa5968

[52] Lohbeck M, Poorter L, Martínez-Ramos M, et al. Biomass is the main driver of changes in ecosystem process rates during tropical forest succession. Ecology. 2015;**96**(5):1242-1252. DOI: 10.1890/14-0472.1

[53] Arroyo-Rodriguez V, Melo FPL, Martinez-Ramos M, et al. Multiple successional pathways in human-modified tropical landscapes: New insights from forest succession, forest fragmentation and landscape ecology research. Biological Reviews. 2015;**92**(1):326-340. DOI: 10.1111/brv.12231

[54] Laurance WF, Camargo JL, Fearnside PM, et al. An Amazonian rainforest and its fragments as a laboratory of global change. Biological Reviews. 2017. DOI: 10.1111/brv.12343

[55] Costa MH, Pires GF. Effects of Amazon and Central Brazil deforestation scenarios on the duration of the dry season in the arc of deforestation. International Journal of Climatology. 2010;**30**(13):1970-1979. DOI: 10.1002/joc.2048

[56] Choat B, Jansen S, Brodribb T, et al. Global convergence in the vulnerability of forests to drought. Nature. 2012;**491**(August):752-755. DOI: 10.1038/nature11688

[57] Santos MG, Oliveira MT, Figueiredo KV, et al. Caatinga, the Brazilian dry tropical forest: Can it tolerate climate changes? Theoretical and Experimental Plant Physiology. 2014;**26**(1):83-99. DOI: 10.1007/s40626-014-0008-0

[58] Vieilledent G, Gardi O, Grinand C, et al. Bioclimatic envelope models predict a decrease in tropical forest carbon stocks with climate change in Madagascar. Journal of Ecology. 2016;**140**(3):703-715. DOI: 10.1111/1365-2745.12548

[59] Strassburg BBN, Latawiec A, Balmford A. Urgent action on Cerrado extinctions. Nature. 2016;**540**:199

[60] Hobbs RJ, Higgs E, Harris JA. Novel ecosystems: Implications for conservation and restoration. Trends in Ecology and Evolution. 2009;**24**(11):599-605. DOI: 10.1016/j.tree.2009.05.012

Chapter 5

Advances in the Knowledge of the Vegetation of Hispaniola (Caribbean Central America)

Ana Cano-Ortiz, Carmelo Maria Musarella,
Carlos José Piñar Fuentes,
Carmen Bartolomé Esteban, Ricardo Quinto-Canas,
Carlos José Pinto Gomes, Sara del Río and
Eusebio Cano

Additional information is available at the end of the chapter

http://dx.doi.org/10.5772/intechopen.72090

Abstract

The vegetation types and floristic diversity in the Dominican Republic are analysed, a territory with a tropical climate and ombrotypes that range from dry to humid-hyperhumid, due to the Atlantic winds and the phenomenon known as rain shadows. The presence of high mountains and different substrates have led to a rich flora, and as a result, a high diversity of habitats, among which two large forest types are particularly notable: (1) the dry forest with 81 endemic species, of which 10 are trees, 65 shrubs, 5 climbers and 1 herbaceous species, and an absence of epiphytes and (2) the cloud forest with 19 trees, 20 shrubs, 8 climbers, 4 epiphytes, and 6 herbaceous species. In all cases, these plant communities are regarded as endemic due to their high rate of endemic species. In spite of their importance for conservation, these habitats are highly deteriorated due to deforestation for agriculture, to obtain timber, and even to add to tourism infrastructures.

Keywords: endemics, conservation, deterioration, diversity, forest, structure

1. Introduction

The analysis of the vegetation of Hispaniola reveals different situations in terms of their state of conservation, the floristic diversity of the communities and habitat types. The process observed in the study territory resembles the reports by other authors for other countries: namely that agriculture and tourism are the primary factors affecting conservation. The island of Hispaniola has a wide floristic diversity with over 6000 endemic plants, and a high rate of plant communities. This is due to several factors such as orography, edaphology, and its strategic situation in the central Caribbean, which has meant that species have arrived through migratory routes from North and South America. The northernmost areas of the island, exposed to the Atlantic Ocean, are rainier than those in the Caribbean Sea (**Figure 1**).

Some works on vegetation published in the local journal of the Santo Domingo Botanical Garden (Moscoso) have been of assistance in this research. Deforestation changes the structure of the cloud forest [1] and gives rise to a "calimetal" of *Gleichenia bifia* (Willd.) Spreng. and *Dicranopteris flexuosa* (Schrad.) Underw., whose vigorous growth and dense coverage hinder the regeneration of the forest's own dynamic stages of deforestation in tropical areas.

The deterioration of the forests on the island of Hispaniola is exacerbated by the Caribbean hurricanes, which have a serious impact on their structure [2]. However, the mountains on the island are still home to a number of well-conserved areas, thanks to a close linear correlation between environmental factors and forest types [3], and to indigenous people who exploit the forests sustainably, using the wood to build houses and coal using knowledge acquired through tradition [4–6]. The presence of 28 climbing species in the cloud forest versus 44 in the dry forest is due to environmental factors, vegetation structure [7], and deforestation caused by human.

These forests are crucial to maintaining the biodiversity of this island—considered a biodiversity hotspot [8]—, and also for preserving the relation between forest stands and the water cycle, which is of such overriding importance to society [9–11].

The fundamental aim of this chapter is to explain the diversity of plant communities, their causes and their floristic diversity, and highlight their conservation status.

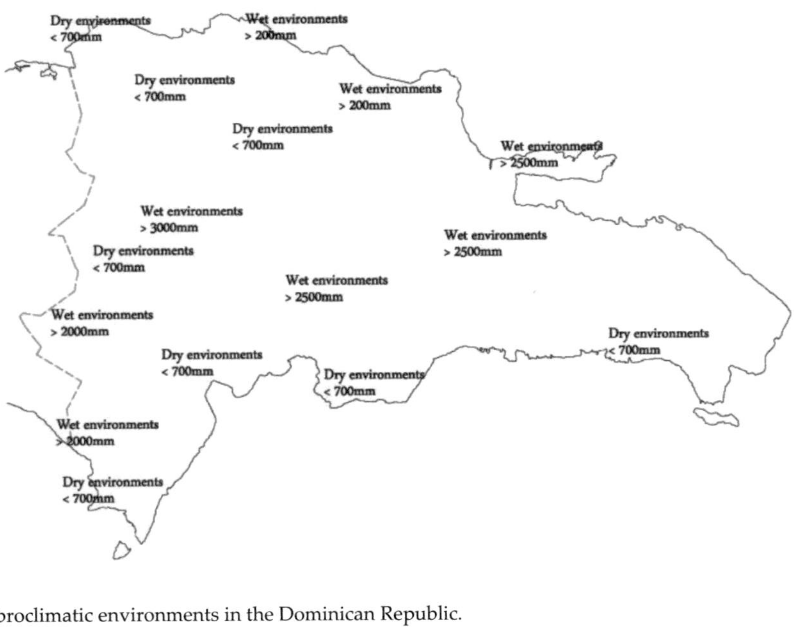

Figure 1. Ombroclimatic environments in the Dominican Republic.

2. Materials and methods

A vegetation study was carried out on the island of Hispaniola, with particular reference to the forests in the Dominican Republic. Sixty-five ecologically and physiognomically homogeneous plots were selected, and the species present on the plots were noted and assigned a phytosociological abundance-dominance index; this index was subsequently converted to the Van der Maarel (1979) index [12]. We used our previous work as a basis for the bioclimatic and biogeographical classification of the territory [13, 14].

A statistical treatment was applied to obtain cluster and PCA. To separate the two major vegetation units (cloud forest and dry forest), we conducted a regression analysis (ANOVA) for the species *Magnolia hamorii* Howard (Maha), *Sideroxylon salicifolium* (L.) Sw. (Sisa), *Metopium toxiferum* (L.) Krug. & Urb. (Meto) and *Sideroxylon foetidissimum* Jacq. (Sifo); and a canonical correspondence analysis (CCA) between the Io (ombrothermic index) and ETP (potential evapotranspiration) of the 65 plots in the sample and the dominant species, located at altitudes below 1,200 m: *Hyeronima domingensis* Urb.(Hydo), *Magnolia hamorii* (Maha), *Magnolia pallescens* Urb. & Ekm.(Mapa) and *Pinus caribaea* Morelet. This last species forms mixed stands with the cloud forest and gives the name to the vegetation class *Byrsonimo-Pinetea caribaea* Samek and Borhidi in Borhidi et al. (1979), a native Caribbean species of subtropical and tropical areas [15]. Outside its distribution area (Central America), this tropical plant acts as an exotic, and is a well adapted in other tropical areas such as southern China [16].

3. Results

The results of the ANOVA regression analysis reveal the close correlation between the species present in the forests and the potential evapotranspiration and Io index. The species in the cloud forest are closely correlated with the ombroclimatic index, and in the dry forest with potential evapotranspiration, as shown in **Figures 2** and **3** where R2 always higher than 0.9.

The results of the regression analysis are Maha (*Magnolia hamorii*): $y = 0.08585x - 1.5$, $R^2 = -0951$; Sisa (*Sideroxylon salicifolium*): $y = 0.5799x - 0.4$, $R^2 = 0.9994$; Meto (*Metopium toxiferum*): $y = 0.4065x - 0.5$, $R^2 = 0.9979$; Sifo (*Sideroxylon foetidissimum*): $y = 0633x - 0.5$, $R^2 = 0.9708$; highlighting the clear correlation between the abundance values for these species and the values for Io and ETP (ANOVA) (**Figure 2**).

The CCA analysis shows two groups of inventories: R1-R39, representing the dry forest and dominated by *Bursera simaruba* (L.) Sarg., *Metopium toxiferum* (L.) Krug. & Urb., *Pilosocereus polygonus* (Lam.) B. & R., *Sideroxylon foetidissimum* Jacq., *Sideroxylon salicifolium* (L.) Sw., *Prosopis juliflora* L., *Lemaireocereus hystrix* Britt & Rose, and *Acacia macracantha* H. & B. Ex Willd.; and the group C40–C65 dominated by *Didymopanax tremulus* Krug. & Urb., *Hyeronima domingensis* Urb., *Magnolia hamorii* Howard, *Magnolia pallescens* Urb. & Ekm., *Prestoea montana* (Grah.) Nichol., *Alchornea latifolia* Sw., *Cyathea arborea* (L.) J.E. Smith, and *Cyathea furfuracea* Baker. (**Figure 3**).

The PCA analysis of 28 dominant species produces 2 groups such as A (cloud forest) and B (dry forest) (**Figure 4**). The two groups A and B in the PCA are maintained in the cluster.

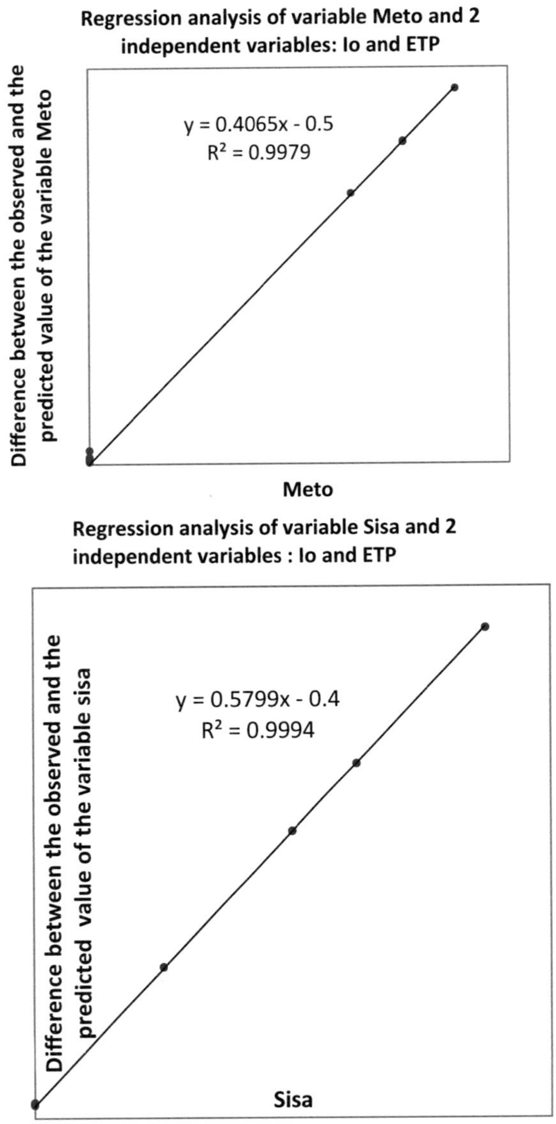

Figure 2. Regression analysis: Meto, Sisa and Io, ETP.

The forests in group B are separated into two subgroups such as G1 and G2. G1 (R1–R18) represents the communities described for the subhumid and humid territories in which the substrate takes prevalence and allows the entry of dry forest species. G2 (R19–R39) is classified into four plant communities.

Subgroup $G2_1$ includes the inventories R19–R48 taken in the Cibao Valley (Monte Cristi), in infratropical environments with a semiarid ombrotype on quaternary loans, with 17 endemic species, *Karwinskia coloneura* Urb. and *Justicia abeggii* Urb. & Ekm., exclusive to the territory. The subgroup is dominated by *Prosopis juliflora* L., *Lemaireocereus hystrix* Britt. & Rose, *Harrisia nashii* Britt. & Rose, *Consolea moniliformis* (L.) Haworth in Steud, and *Pilosocereus polygonus* (Lam.) B. & R.

http://dx.doi.org/10.5772/intechopen.72090

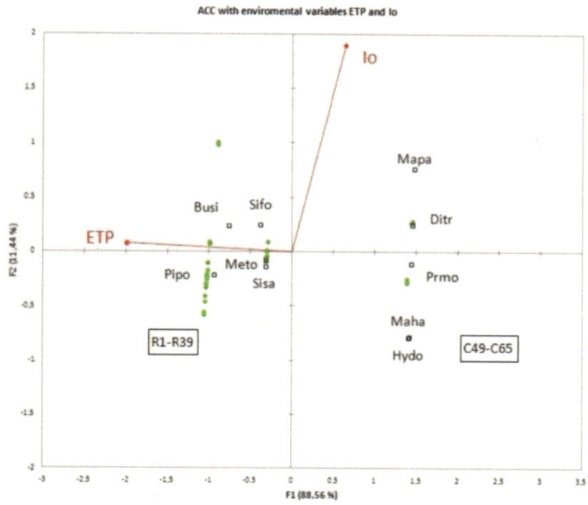

Figure 3. Canonical correspondence analysis (CCA).

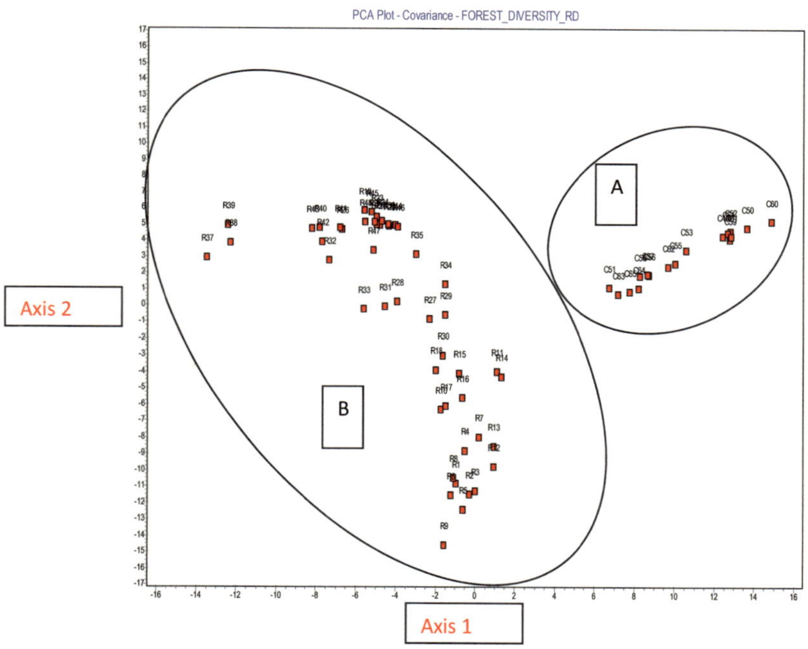

Figure 4. Principal component analysis (PCA) for dry and cloud forest.

Subgroup $G2_2$ includes the inventories R27–R33, a community that grows in the outlying areas of the Cibao Valley in the infra and dry thermotropical thermotype. This is an intricate forest dominated by *Erythroxylum rotundifolium* Lunan and *Maytenus buxifolia* (A. Rich.) Griseb., with 36 endemic plants, of which the following are exclusive to the territory: *Lantana leonardorum* Moldenke, *Lantana pauciflora* Urb., *Croton gonaivensis* Urb. & Ekm., *Lantana buchii* Urb., *Guettarda tortuensis* Urb. & Ekm., and *Galactia synandra* Urb. (**Figure 5**).

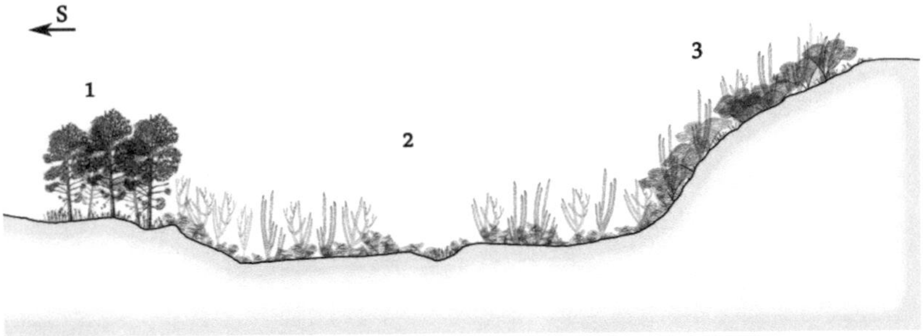

Figure 5. Profile of the vegetation of the dry forest of Cibao Valley. 1. *Leptogono buchii-Pinetum occidentalis*. 2. *Harrisio nashii-Prosopidetum juliflorae* 3. *Crotono poitaei-Erythroxyletum rotundifolii* [17].

Subgroup $G2_3$ (R40–R43) is located in areas of Ázua, Bani, San Juan, Enriquillo and Port au Prince (Haiti), which have a dry infratropical thermotype and are home to a spinescent forest growing on Miocene loams. In this forest, there is a predominance of *Guaiacum officinale* L., *Capparis cynophallophora* L., *Bourreria divaricata* (DC.) G. Don and *Cylindropuntia caribaea* (Britt. & Rose) Kunth; and 16 endemic plants, including particularly *Coccothrinax spissa* Bailey, and *Neoabbottia paniculata* (Lam.) Britt. & Rose.

$G2_4$ contains the inventories R37–R39 taken in the dry forest between Pedernales and Barahona. The territory has a semiarid infratropical thermotype and reef limestone substrates. This forest is rich in coarse and spiny plants such as *Cylindropuntia caribaea* (Britt. & Rose) Kunth, *Cameraria linearifolia* Urb., *Bursera simaruba* (L.) Sarg. and *Agave antillarum* Descourt.; and has 24 endemics, including particularly *Melocactus pedernalensis* M. Mejía & R. García, *Malpigina micropetala* Urb., *Thouinia domingensis* Urb. & Radlk., *Thouinidium inaequilaterum* Alain, and *Lonchocarpus pycnophyllus* Urb. These samples are clearly separated from the rest due to the fact that they constitute specific endemic habitats (**Figures 6** and **7** and **Photo 1**).

The type A forest group is located in the northern areas of the great mountains; Cordillera Central, Northern, Bahoruco, Oriental, Los Haitises, all of which have an ombrotype that oscillates between the subhumid and the hyperhumid, and a thermotropical to supratropical thermotype (**Photos 2–4**).

The territory of the Dominican Republic, including some small adjacent islands, covers 48,198 km^2 and accounts for over two-thirds of the territory of the island of Hispaniola, located between parallels 17 and 19°N, and part of the Greater Antilles. The main aim of this chapter is to determine the forest vegetation (cloud forest), floristic diversity, and state of conservation in areas with high rainfall in the Dominican Republic. Most of the botanical studies to date are floristic in nature, and include the studies on the Sierra de Bahoruco, which highlight the substantial rainfall of up to 4000 mm and the very high rate of endemic species. Studies by various authors in the central, northern, and eastern mountain ranges containing the cloud forest [18–24], and which together with our own previous studies, form the basis of this present work [25–35]. All the above-mentioned studies focus on the knowledge of the

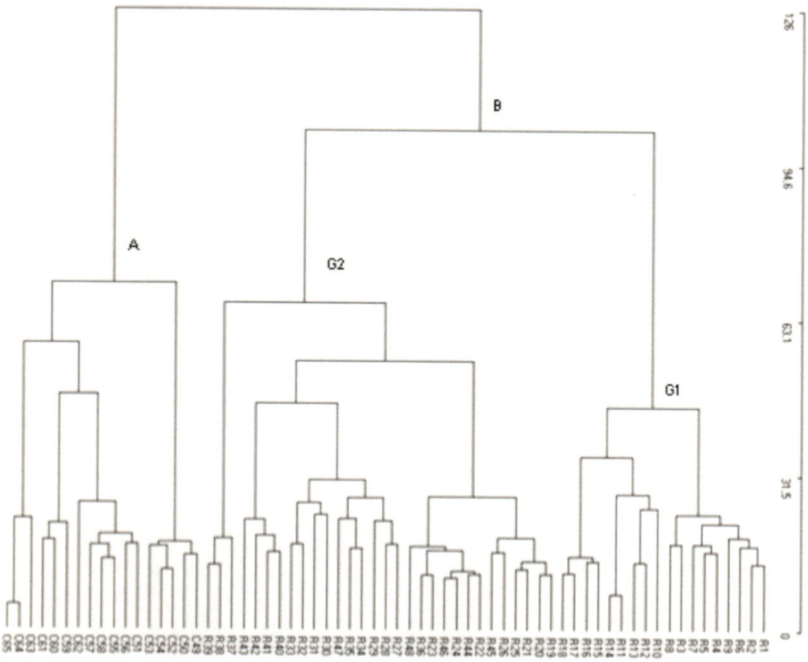

Figure 6. Cluster analysis. Diversity of forest types.

Figure 7. Profile of the vegetation of the cloud forest of Sierra Bahoruco-Barahona-Pedernales 1. *Rhabdadenia biflora-Laguncularietum racemosae* and *Lonchocarpo pycnophylli-Conocarpetum erecti*. 2. Salt marshes of *Batidi-Salicornietea*. 3. *Lonchocarpo pycnophylli-Cylindropundietum caribaeae*. 4. *Melocacto pedernalensis-Leptochloopsietum virgatae*. 5. Broad-leaved forest. 6. Cloud forest of *Prestoea montana* [17].

flora, with passing references to the vegetation; herein lies the interest of this work. The floristic approach is important as these are biodiversity hotspots [36], with a 33% rate of endemics compared to the total flora.

Photo 1. Dry forest of *Pilosocereus polygonus* (Cibao Valley, Dominican Republic).

Photo 2. Subtropical forests of *Pinus occidentalis* in the Cordillera Central, and the heath grass formation of *Danthonia domingensis*.

Photo 3. Meso and supratropical cloud forests in the Cordillera Central.

In the analysis of the diversity of the 2 forest types (dry and rain); in the dry forest, there are 61 tree species and 5 stipes, of which 10 are endemic; compared to 75 in the cloud forest and 4 stipes, of which 19 are endemic. There is more number of epiphytes in the cloud forest, with

Photo 4. Vegetation of Los Haitises (A6). 1. Latifoliate mahogany forest *Swietenia mahagoni*. 2. *Ormosia krugii* and *Prestoea montana* which is enriched on the summits with *Didymopanax morototoni*.

	Trees	Stipes	Shrubby	Succulents	Climbers	Epiphytes	Herbaceous	T. species
Dry forest	61	5	158	16	44	13	28	322
Rain forest	75	4	63	0	28	36	43	244

Table 1. Species diversity in dry and cloud forests in the Dominican Republic.

36 species, than in the dry forest, with 13 species (**Table 1**). Shannon's index gives values of 3384 and 3528 for the dry and cloud forest, respectively.

4. Discussion and conclusions

There are areas with dry forest plant cover in the Dominican Republic; these are located in the southwest of the island, specifically in Procurrente de Barahona, Azua-Valle de San Juan-Hoya Enriquillo and in its extension toward western Haiti, including Beata Island and the Cibao Valley, Monte Cristi. These territories have a high rate of endemics, therefore, the habitats where they grow should also be considered endemic. The cloud forest, however, is always found in the mountains as a consequence of the moisture-laden winds from the Atlantic. In both cases, the forest distribution responds to the environmental factors on the island, which has a high rate of endemics owing to the island phenomenon, making it a hotspot in the Caribbean.

The presence of two groups of forests due to the humidity gradient [37] in the Dominican Republic, which varies between semiarid and hyperhumid. This classification is based on the floristic composition and less on forest structure [38]. The cloud forest in Ebony Green (Cordillera Central) has suffered frequent fires, causing an invasion of *Gleichenia bifida* and *Dicranopteris pectinata*, which prevent the germination and development of primary forest species and in turn leads to a loss of seeds and intensification in the rate of invasive species. This situation requires rapid human intervention [39] (**Photo 5**), which should be aimed at reinforcing the seed bank of species of interest to conservation in botanical gardens in the Caribbean

Photo 5. Deforested area in the central range invaded by *Gleichenia bifida* (Dominican Republic).

Photo 6. Deforestation of the dry forest in the Cibao Valley (Dominican Republic).

[40]. The rain forests on the ridges and mountain ranges of Hispaniola are rich in *Cecropia schreberiana* Miq. [41] and species of the genera *Magnolia, Prestoea, Didymopanax, Cyathea*; the greatest botanical diversity is found in the oldest mountains, while the younger formations are less floristically diverse [42]. The diversity of trees and epiphytes is lower in the dry forest than in the cloud forest, but there are more succulents from the Cactaceae and Agavaceae family; we found no interactions between these families and parasitic elements in the Dominican Republic as occur in Puerto Rico. This type of dry forest dominated by succulent species has a low rate of epiphytes (4%) compared to 15% for the cloud forest. Epiphytes are renowned for their abundance in dry forests: *Tillandsia recurvata* (L.) L, *Broughtonia domingensis* (Lindl.) Rolf, *Tillandsia balbisiana* Schultes, *Tillandsia usneoides* (L.) L, and *Bromelia pinguin* L.

Deforestation caused by fire or heavy machinery is causing changes in the landscape, leading to the loss of species and plant cover (**Photo 6**) in areas where there is insufficient knowledge of the flora and vegetation, despite the fact that, so far, we have described nine phytosociological classes and more than 20 associations and plant communities on the island.

Global syntaxonomical checklist of the known vegetal diversity of Hispaniola (syntaxonomical approach to Hispaniola)

Number of classes, orders, alliances and associations recognized so far. Classes: 9; Orders: 11; Alliances: 13; Associations: 29

CHRYSOBALANO-ANNONETEA GLABRAE Borhidi & Muñiz in Borhidi, Muñiz & Del-Risco 1979

Tabebuio-Bucidetalia (Lvov 1967) Borhidi & Del-Risco in Borhidi, Muñiz & Del-Risco 1979

Marcgravio rubrae-Pterocarpion officinalis Cano, Velóz, Cano-Ortiz et Esteban Ruiz 2009

1-Roystoneo hispaniolanae-Pterocarpetum officinalis Cano, Velóz, Cano-Ortiz et Esteban Ruiz 2009

BYRSONIMO-PINETEA CARIBAE Samek and Borhid in Borhidi et al. 1979

Pinetalia occidentalis-maestrensis Knapp 1964 in Borhidi et al. 1979

Ilici tuerckheimi-Pinion occidentalis Cano, Velóz et Cano-Ortiz 2011

1-Dendropemon phycnophylli-Pinetum occidentalis Cano, Velóz et Cano-Ortiz 2011

2-Cocotrino scopari-Pinetum occidentalis Cano, Velóz et Cano-Ortiz 2011

RHIZOPHORO-AVICENNIETEA GERMINANTIS Knapp (1964) em. Borhidi & Del-Risco in Borhidi et al. 1979

Rhizophoretalia Cuatrecasas 1958

Al. Dalbergio-Rhizophorion manglis (Borhidi 1991) Cano, Cano-Ortiz, Velóz, Alatorre et Otero 2012

1-Machario lunati-Rhizophoretum manglis Cano, Cano-Ortiz & Velóz ex Cano, Cano-Ortiz, Velóz, Alatorre et Otero 2012

Avicennietalia germinantis Cuatrecasas 1958

Conocarpo-Laguncurion racemosae Cuatrecasas 1958

2-Batidi-Avicennietum germinantis Borhidi & Del-Risco & Borhidi 1991

(syn. As. Laguncurio racemosae-Avicennietum germinantis Reyes & Acosta 2003; As. Avicennietum germinantis Reyes & Acosta 2003)

3-Rhabdadenio biflorae-Laguncularietum racemosae Cano, Cano-Ortiz & Velóz ex Cano, Cano-Ortiz, Velóz, Alatorre et Otero 2012

4-Conocarpo erectae-Coccolobetum uviferae Reyes in Reyes & Acosta 2003

(syn. Conocarpetum erectae Reyes in Reyes & Acosta 2003)

5-Sthalio monospermae-Laguncularietum racemosae Cano, Cano-Ortiz & Velóz ex Cano, Cano-Ortiz, Velóz, Alatorre et Otero 2012

6-Lonchocarpo pycnifolii-Conocarpetum erecti Cano, Cano-Ortiz & Velóz ex Cano, Cano-Ortiz, Velóz, Alatorre et Otero 2012

TABEBUIO-BURSERETEA Knapp (1964) Borhidi 1991

Tabebuio-Burseretalia Knapp (1964) Borhidi 1991

Leptogono buchii-Tabebuion berterii Cano, Cano-Ortiz, del Río, Velóz et Esteban Ruíz 2014

1-Coccotrino argentei-Tabebuietum berterii Cano, Cano-Ortiz, del Río, Velóz et Esteban Ruíz 2014

2-Zombio antillari-Leptogonetum buchii Cano, Cano-Ortiz, del Río, Velóz et Esteban Ruíz 2014

Calliandro haematommae-Phyllanthion nummularioidis Cano, Cano-Ortiz, del Río, Velóz et Esteban Ruíz 2014

3-Garcinio glaucescentis-Phyllanthetum numularioidis Cano, Cano-Ortiz, del Río, Velóz et Esteban Ruíz 2014

4-Tabebuio ophiolithicae-Randietum aculeati Cano, Cano-Ortiz, del Río, Velóz et Esteban Ruíz 2014

PHYLLANTHO-NEOBRACETEA VALENZUELANAE Borhidi & Muñiz in Borhidi et al. 1979

Ariadno-Phyllanthetalia Borhidi & Muniz in Borhidi et al. 1979

Tetramicro canaliculatae-Leptochloopsion virgatae Cano, Velóz et Cano-Ortiz 2010

1-Leptogono buchii-Leptochloopsietum virgatae Cano, Velóz et Cano-Ortiz 2010

Rondeletio christii-Pinion occidentalis Cano, Cano-Ortiz, del Río, Velóz et Esteban Ruíz 2014

2-Leptogono buchii-Pinetum occidentalis Cano, Veloz & Cano Ortiz 2011

COCCOTHRINACETO-PLUMERIETEA Knapp in Boirhi 1991

Lantano-Cordietalia Borhidi in Borhidi et al. 1979

Crotono poitaei-Leptochloopsion virgatae Cano, Velóz et Cano-Ortiz 2010

1-Crotono astrophori-Leptochloopsietum virgatae Cano, Velóz et Cano-Ortiz 2010

2-Melocacto pedernalensis-Leoptochloopsietum virgatae Cano, Velóz et Cano-Ortiz 2010

3-Solano microphylli-Leptochloopsietum virgatae Cano, Velóz et Cano-Ortiz 2010

Eugenio-Metopietalia toxiferi Knapp (1942) Borhidi 1991

Eugenio-Capparidion Borhidi in Borhidi et al. 1959

4-Chrysophyllo oliviformi-Sideroxyletum salicifolii Cano & Velóz 2012

5-Zamio debilis-Metopietum toxiferi Cano & Velóz 2012

6-Coccotrino gracili-Burseretum simarubae Cano, Cano-Ortiz et Velóz 2015

CERCIDI-CEREETEA Borhidi 1996

Ritterocereetalia hystricis Borhidi 1996

Harrio nashii–Acacion skleroxylae Cano, Cano-Ortiz & Velóz ex Cano-Ortiz, Musarella, Spampinato, Velóz et Cano 2015

1-Harrisio nashii–Prosopidetum juliflorae Cano, Cano-Ortiz & Velóz ex Cano-Ortiz, Musarella, Spampinato, Velóz et Cano 2015

2-Crotono poitaei–Erythroxyletum rotundifolii Cano, Cano-Ortiz & Velóz ex Cano-Ortiz, Musarella, Spampinato, Velóz et Cano 2015

3-Lonchocarpo pycnophylli–Cylindropuntietum caribaeae Cano, Cano-Ortiz &Velóz ex Cano-Ortiz, Musarella, Spampinato, Velóz et Cano 2015

4-Neoabbottio paniculatae–Guaiacetum officinalis Cano, Cano-Ortiz & Velóz ex Cano-Ortiz, Musarella, Spampinato, Velóz et Cano 2015

WEINMANNIO-CYRILLETEA Knapp 1964

Weinmannio-Cyrilletalia Knapp 1964

1.-Community of Cyathea furfuracea and Prestoea motana

2.-Community of Ormosia krugii and Prestoea montana

OCOTEO-MAGNOLIETEA Borhidi and Muñiz in Borhdi et al. 1979

Ocoteo-Magnolietalia Muñiz in Borhdi et al. 1979

1.-Community of Hyeronima Montana and Magnoliea pallescens

2.-Community of Hyeronima dominguensis and Magnolia hamorii

Acknowledgements

We would like to thank Ms Pru Brooke Turner (MA Cantab) for the English translation of this chapter. Some paragraphs have been taken from the previous publication of Cano-Ortiz et al. [35], having previously obtained permission from the same.

Author details

Ana Cano-Ortiz[1], Carmelo Maria Musarella[2], Carlos José Piñar Fuentes[1], Carmen Bartolomé Esteban[3], Ricardo Quinto-Canas[4], Carlos José Pinto Gomes[5], Sara del Río[6] and Eusebio Cano[1*]

*Address all correspondence to: ecano@ujaen.es

1 Dept. of Animal and Plant Biology and Ecology, Botany Department, University of Jaén, Spain

2 Dept. di AGRARIA, Università "Mediterranea" di R. Calabria, Località Feo di Vito, Calabria, Italy

3 Dept. Life Sciences, Universidad de Alcalá, Madrid, Spain

4 CCMAR—Centro de Ciências do Mar, University of Algarve, Portugal

5 Dept. de Paisagem, Ambiente e Ordenamento / Instituto de Ciências Agrárias e Ambientais Mediterrânicas (ICAAM), Universidade de Évora, Portugal

6 Department of Biodiversity and Environmental Management (Botany), Faculty of Biological and Environmental Sciences, University of León, León, Spain

References

[1] Oliveras I, Malhi Y, Salinas N, Huaman V, Urquiaga-Flores E, Kala-Mamani J, Quintano-Loaiza JA, Cuba-Torres F, Lizarraga-Morales N, Roman-Cuesta RM. Changes in forest structure and composition after fire in tropical monatane cloud forests near the Andean treeline. Plant Ecology & Diversity. 2014;**7**(1-2):329-340

[2] Saito S. Effects of severe typhoon on forest dynamics in a warn-temperature evergreen broad-leaved forest in Southwestern Japan. Journal of Forestry Research. 2002;**7**: 137-143

[3] Phua M, Minowa M. Evaluation of environmental functions of tropical forest in Kinabalu Park, Sabah, Malaysia using GIS and remote sensing techniques: Implications to forest conservation planning. Journal of Forestry Research. 2000;**5**:123-131

[4] Berkes F. Indigenous ways of knowing and the study of environmental change. Journal of the Royal Society of New Zealand. 2009;**39**(4):151-156

[5] Agrawal A. Why indigenous knowledge? Journal of the Royal Society of New Zealand. 2009;**39**(4):157-158

[6] Robson JP, Miller AM, Idrobo CJ, Burlando C, Deutsch N, Kocho-Schellenberg JE, Pengelly D, Turner KL. Building communities of learning: Indigenous ways of knowing in contemporary natural resources and environmental management. Journal of the Royal Society of New Zealand. 2009;**39**(4):173-177

[7] Almeida de Oliveira E, Schwantes Marimon B, Feldpausch TR, Rinaldi Colli G, Marimon-Junior BH, Lloy J, Lenza E, Maracahipes L, Oliveira-Santos C, Oliver LP. Diversity, abundance and distribution of lianas of the cerrado-Amazonian Forest transition, Brazil. Plant Ecology & Diversity. 2014;**7**(1-2):231-240

[8] Maunder M, Leiva A, Santiago-Valentín E, Stevenson DW, Acevedo-Rodríguez P, Meerow AW, Mejía M, Clubbe C, Francisco-Ortega J. Plant conservation in the Caribbean Island biodiversity hotspot. Botanical Review. 2008;**74**:197-207

[9] Nakashizuka T. The role of biodiversity in Asian forests. Journal of Forestry Research. 2004;**9**:293-298

[10] Kaimowitz D. Forest and water: A polycy perspective. Journal of Forestry Research. 2004;**9**:289-291

[11] Uzawa H. Forest and social common capital. Journal of Forestry Research. 2004;**9**:283-288

[12] Van Der Maarel E. Transformation of cover-abundance values in phytosociology and its effects on community similarity. Vegetatio. 1979;**39**:97-114

[13] Cano E, Cano Ortiz A, Del Río S, Alatorre J, Veloz A. Bioclimatic map of the Dominican Republic. Plant Sociology. 2012;**49**:81-90

[14] Cano E, Cano Ortiz A. Establishment of Biogeographic Areas by Distributing Endemic Flora and Habitats (Dominican Republic, Haiti R.). In Global Advances in Biogeography. Croatia: Ed. Lawrence Stevens INTECH; 2012. pp. 99-118

[15] Nobis MP, Traiser C, Roth-Nebelsick A. Latitudinal variation in morphological traits of the genus Pinus and its relation to environmental and phylogenetic signals. Plant Ecology & Diversity. 2012;**5**(1):1-11

[16] Wei Y, Ouyang Z, Miao H, Zheng H. Exotic *Pinus caribaea* causes soil quality to deteriórate on former abandoned land compared to an indigenous *Podocarpus* plantation in the tropical forest area of southern China. Journal of Forestry Research. 2009;**14**:221-228

[17] Cano Ortiz A, Musarella CM, Cano E. Biogeografical areas of Hispaniola (Dominican Republic, Republic of Haiti) in Ecology. Intech Open Science. 2017:165-189

[18] May TH, Pequero B. Vegetación y flora de la Loma el Mogote, Jarabacoa, Cordillera Central, República Dominicana. Moscosoa. 2000;**11**:11-37

[19] Mejía M, García R, Jiménez F. Sub-región fitogeográfica Barbacoa-Casabito: riqueza florística y su importancia en la conservación de la flora de la Isla Española. Moscosoa. 2000;**11**:57-106

[20] May TH. Fases tempranas de la sucesión en un bosque nublado de Magnolia pallescens después de un incendio (Loma de Casabito, Reserva Científica Ébano Verde, Cordillera Central, República Dominicana). Moscosoa. 1997;**9**:117-144

[21] Zanoni T, Mejía M, Pimentel JD, García R. Flora y Vegetación de los Haitises República Dominicana. Moscosoa. 1990;**6**:46-98

[22] Mejía M, Jiménez F. Flora y Vegetación de la Loma Humeadora, Cordillera Central, República Dominicana. Moscosoa. 1998;**10**:10-46

[23] Hager J, Zanoni T. La vegetación natural de la República Dominicana. Una nueva clasificación. Moscosoa. 1993;**7**:39-81

[24] Höner D, Jímenez F. Flora vascular y vegetación de la Loma La Herradura (Cordillera Oriental, República Dominicana). Moscosoa. 1994;**8**:65-85

[25] Cano E, Veloz Ramírez A, Cano Ortiz A, Esteban FJ. Distribution of Central American Malastomataceae: Biogeographical Analysis of the Caribbean Islands. Acta Botanica Gallica. 2009;**156**(4):527-557

[26] Cano E, Veloz Ramírez A, Cano Ortiz A, Esteban FJ. Analysis of the Pterocarpus officinalis forests in the Gran Estero (Dominican Republic). Acta Botanica Gallica. 2009; **156**(4):559-570

[27] Cano E, Veloz A, Cano Ortiz A. Contribution to the biogeography of the Hispaniola (Dominican Republic, Haiti). Acta Botanica Gallica. 2010;**157**(4):581-598

[28] Cano E, Veloz A, Cano Ortiz A. The habitats of Leptochloopsis virgata in the Dominican Republic. Acta Botanica Gallica. 2010;**157**(4):645-658

[29] Cano E, Veloz Ramírez A, Cano Ortiz A. Phytosociological study of the *Pinus occidentalis* woods in the Dominican Republic. Plant Biosystems. 2011;**145**(2):286-297

[30] Cano E, Veloz A, Cano Ortiz A, Alatorre J, Otero R. Comparative analysis of the mangrove swamps of the Caribbean and those of the state of de Guerrero. Mexico. Plant Biosystems. 2012;**146**(suplemento 1):112-130

[31] Cano E, Cano Ortiz A, del Río S, Veloz A, Esteban Ruíz FJ. A phytosociological survey of some serpentine plant communities in the Dominican Republic. Plant Biosystems. 2013;**148**(2):200-212

[32] Cano E, Veloz A, Cano Ortiz A. Rain forests in subtropical mountains of Dominican Republic. American Journal of Plant Science. 2014;**5**:1459-1466

[33] Cano E, Cano Ortiz A, Veloz A. Contribution to the knowledge of the edaphoxerophilous communities of the Samana peninsula (Dominican Republic). Plant Sociology. 2015; **52**(1):3-8

[34] Cano Ortiz A, Musarella CM, Piñar JC, Veloz A, Cano E. The dry forest in the Dominican Republic. Plant Biosystems. 2015;**149**(3):451-472

[35] Cano Ortiz A, Musarella CM, Piñar Fuentes JC, Pinto Gomes CJ, Cano E. Forests and landscapes of Dominican Republic. British Journal of Applied Science & Technology. 2015;**9**(3):231-242

[36] Cano Ortiz A, Musarella CM, Piñar Fuentes JC, Pinto Gomes CJ, Cano E. Distribution patterns of endemic flora to define hotspots on Hispaniola. Systematics and Biodiversity. 2016;**14**(3):261-275

[37] Butt N, Malhi Y, New M, Macia MJ, Lewis SL, Lopez-Gonzalez G, Laurance WF, Laurance S, Luizao R, Andrade A, Baker TR, Almeida S, Oliver L. Shifting dynamics of climate-functional groups in old-growth Amazonian forests. Plant Ecology & Diversity. 2014;**7**(1-2):267-279

[38] Gardener MR, Bustamante RO, Herrera I, Durigan G, Pivello VR, Moro MF, Stoll A, Langdon B, Baruch Z, Rico A, Arredondo-Nuñez A, Flores S. Plant invasions research in Latin America: Fast track to a more focused agenda. Plant Ecology & Diversity. 2012; **5**(2):225-232

[39] Faggi A, da Costa ML, Pereira TS, Balcazar Sol T, Mejía M. Latin America and Caribbean botanic gardens: advances an challenges at national and regional levels. Plant Ecology & Diversity. 2012;**5**(2):259-263

[40] Sheil D, Padmanaba M. Innocent invaders? A preliminary assessment of Cecropia, an American tree, in Java. Plant Ecology & Diversity. 2011;**4**(2-3):279-288

[41] Spehn EM, Rudmann-Maurer K, Körner C. Mountain diversity transitions. Plant Ecology & Diversity. 2011;**6**(1):101-137

[42] Guerrero PC, Carvallo GO, Massar JM, Rojas-Samdoval J, Sanz V, Medel R. Ecology and evolution of negativa and positive interactions in Cactaceae: Lessons and pending tasks. Plant Ecology & Diversity. 2012;**5**(2):205-215

Chapter 6

Vegetation Structure and Prioritizing Plants for Eco-Restoration of Degraded Wildlife Corridor in Dry Tropical Forest of South India

Paramesha Mallegowda, Siddappa Setty and Ganesan Rengaian

Additional information is available at the end of the chapter

http://dx.doi.org/10.5772/intechopen.72706

Abstract

Wildlife corridors are critical to manage wildlife and maintain ecological processes. However, they are fragmented and degraded due to various anthropogenic activities. Fragmentation in turn affects population viability of species by affecting their dispersal, re-colonization and genetic exchanges. But the process can be reversed through restoration and management of 'functional corridors'. So far in the forestry sector, monoculture plantations are known to be the ideal reforestation/afforestation strategy to restore degraded landscape but experts argue that monoculture plantations have failed to recover former biological diversity. Therefore, for successful eco-restoration, first, the regional plant stock has to be identified and then suitable plant species have to be prioritized. The habitat enrichment through assisted vegetation method in the degraded wildlife corridors can improve green cover and also bring back the original vegetation. The study was conducted in the Edeyarahalli-Doddasampige wildlife corridor area, which is part of Biligiri Rangaswamy Temple Tiger Reserve, Western Ghats, India. The vegetation was enumerated through transect and quadrate method. The vegetation structure was analyzed and ten suitable native plant species were prioritized for eco-restoration. The priority was given based on site condition and socio-ecological importance of the plants such as trees with timber value, non-timber forest products, nectar source for honey bees and also food source for elephants. At a time of unprecedented forest destruction, the interventions made through this line of research would not only improve the habitat quality but also increase the functionality of wildlife corridors by providing safe passage for animals' movement. In addition to this, convergence of local multistakeholders and their responsibility needs to be explored toward eco-restoration process.

Keywords: Biligiri Rangaswamy Temple Tiger Reserve, restoration, Western Ghats, wildlife corridor

1. Introduction

The world's tropical forests are being fragmented and degraded with significant loss of species diversity and ecosystem services [1–4]. Unplanned infrastructure development in forest landscapes, clearing of forest land for expansion of human habitation as well as farmland, and unsustainable extraction of forest resources can create growing pressures and also inflict negative impacts on wildlife habitat [5–7]. According to meta-population, meta-community and island-biogeography theories, degradation and fragmentation of natural wildlife habitats could lead to the extinction of many species across the globe due to loss of sub-population connectedness and inbreeding depression [4, 8]. Therefore, at the time of unprecedented wildlife habitat destruction, eco-restoration of degraded forest areas particularly wildlife corridors is gaining global importance and also emerging as a practical conservation strategy [9–12]. Under the 'Green India Mission', the Indian government is planning to double afforestation efforts by 2020 [13] and also planning to buy private plantations to restore elephant corridors [14, 15].

According to the 'Field of Dreams Hypothesis', *if a habitat is successfully restored, the species will return* but we need to refine the appropriate restoration strategy. So far in the forestry sector, monoculture plantations are known to be the ideal reforestation strategy to restore degraded landscapes [16–18] but experts argue that monoculture plantations failed to recover their former biological diversity [19–21]). Therefore, to reverse the effect, the eco-restoration method would be the appropriate strategy. Habitat enrichment through assisted vegetation method can improve green cover as well as bring back the native vegetation and provide resource rich passage for animals' movement. However, as a first step in the eco-restoration activity, the regional plant stock has to be assessed and then suitable native plant species has to be prioritized based on their socio-ecological importance and site condition [22]. In addition to this, the species which are selected for eco-restoration should be strong and hard enough to withstand and survive in the prevailing climatic conditions; mainly heavy rain and dry seasons [16]. This is because, the type of forest occurring naturally in a place is the result of the complex influence of the climatic, edaphic, topographic, and biotic factors of the locality [23].

The Edeyaralli-Doddasampige wildlife corridor (ED corridor) in Biligiri Rangaswamy Temple Tiger Reserve (BRT), Western Ghats is one such biodiversity rich forest landscape but subjected to various land-use practices leading to fragmentation and degradation of wildlife habitat and wildlife migratory routes. Therefore, action and restoration research has been planned in this degraded corridor to maintain the habitat quality and also increase the functionality of the corridor through assisted vegetation enrichment. For successful eco-restoration, first, the regional plant stock has to be identified and then suitable plant species have to be prioritized. In this study, we have addressed the following two research questions; (i) How are the plant community variables such as species richness, density, diversity and IVI (Importance Value Index) distributed among life forms in the corridor landscape?, (ii) How do we prioritize the suitable plant species/categories for eco-restoration of degraded wildlife corridor?

2. Methods

2.1. Study site

The study has been carried out at Edeyarahalli-Doddasampige wildlife corridor (ED corridor), which is one of the degraded but ecologically important functional corridors between Biligiri Rangaswamy Temple Tiger Reserve (BRT) and Malai Mahadeswara Hills Wildlife Sanctuary (MM Hills) (**Figure 1**). The dimension of the ED corridor is 0.5 km in length and 2 km in width and the geographical coordinates are 11°55′15″ to 11°56′15″N and 77°15′20″ to 77°15′45″E. The corridor landscape is largely in the dry deciduous and scrub forest type. It harbors rich floral and faunal diversity, mainly IUCN red listed mammal species such as Asian elephant (*Elephas maximus*), Bengal tiger (*Panthera tigris*), Indian leopard (*Panthera pardus*) and Indian wild dog (*Cuon alpinus*). In addition to this, the corridor landscape is inhabited by *Soligas*, an indigenous tribal community and a few other non-tribal communities.

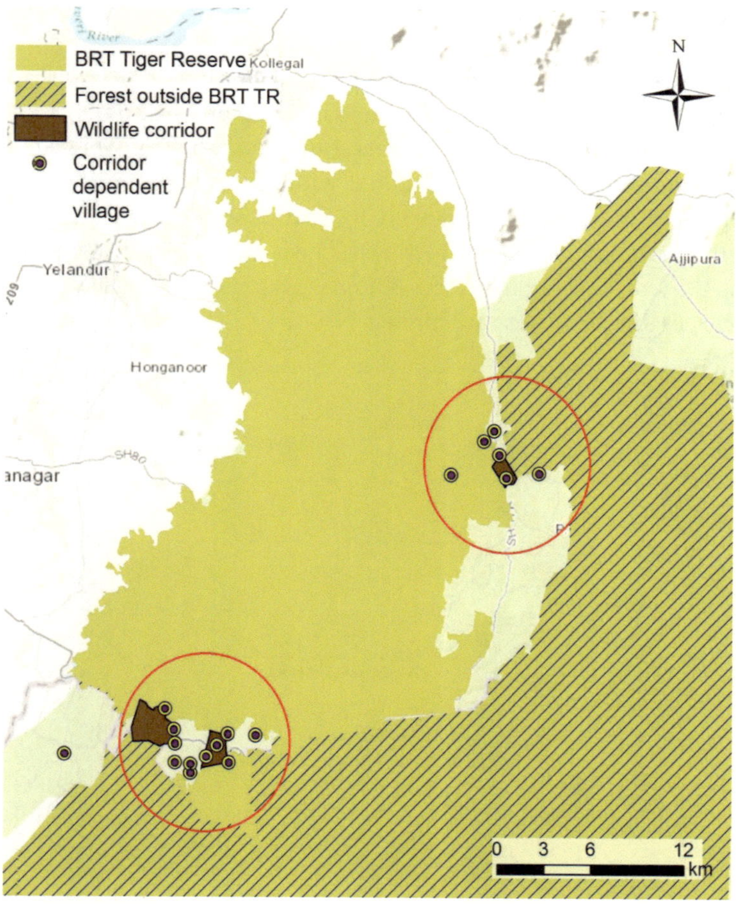

Figure 1. Matrix of forests, wildlife corridors, dependent villages, farmland and road network in and around the corridor landscape (marked in circles).

The corridor landscape is severely degraded due to unplanned land-use practices, past forest management activities- logging and shifting cultivation -and the problem of invasive/exotic plants species [24, 25]. Apart from that, the villagers use this corridor regularly for livestock grazing and fuel wood collection [6]. In addition to this, the state highway (SH-17A) is passing through this wildlife corridor and an average of one vehicle per minute was recorded on this road [26]. This could be an additional threat to the movement of wildlife in this corridor. Irrespective of various threats, ED corridor provides space and passage for more than 15 mammal species (large, medium and small) to move from Western Ghats to forested landscapes of Eastern Ghats [27]. Adjacent to this corridor, in 2007 approx. 25.5 acres of private land was purchased from local farmers to widen the corridor by WTI (Wildlife Trust of India) and its international partner organization International Fund for Animal Welfare (IFAW), with financial support from US Fish and Wildlife Services (USFWS). The land was then handed over to the Karnataka State Forest Department to augment the corridor. This was a pioneering move in corridor conservation in India [6].

2.2. Vegetation enumeration

Transect method was used to enumerate vegetation in the corridor landscape. There were 64 belt transects of 0.1 ha (10 × 100 m), 128 plots of 10 m^2 and 512 plots of 1 m^2 were established to enumerate trees, shrubs and herbaceous plants respectively in the study area (**Figure 2**). Each sampling transect was marked with red ribbons, and the GPS coordinates were recorded at the center of each transect for future study purpose. The sampling was carried out in the month of October, which is the peak wet season in the study area. This is because during the wet season the chances of finding herbaceous species as well as seedlings of woody species in the study area are higher.

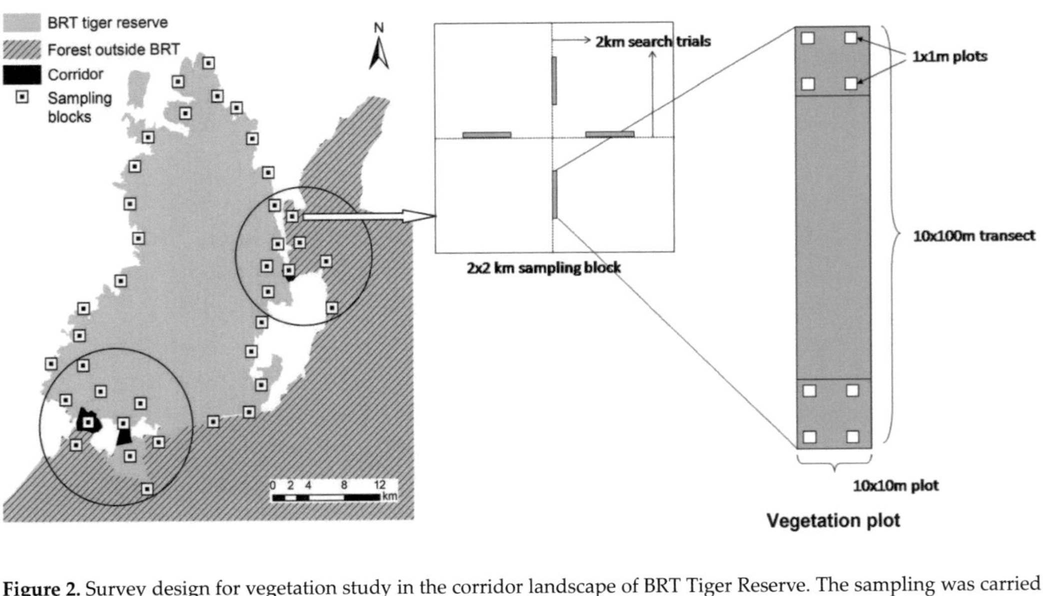

Figure 2. Survey design for vegetation study in the corridor landscape of BRT Tiger Reserve. The sampling was carried out in the blocks which fall within the circles. One 2 × 2 km sampling block consists of four vegetation plots, eight shrub plots and 32 herb plots.

2.2.1. Data collection

In 10 × 100 m transects all stems >5 cm DBH (diameter at breast height - at 130 cm) were enumerated. The DBH of the individual stems were measured for all the species found in the transects using calibrated DBH tape. The height was measured through visual approximation method [28, 29]. In 10 × 10 m plots all the shrubs and saplings of woody plant species whose DBH fell between 1 and 5 cm were counted and named. Finally, in the 1 × 1 m plots all the herbaceous plants and seedlings of woody plant species (whose stem size was <1 cm) were recorded. For most of the species, botanical names and family names were identified and recorded in the field itself. For unidentified plant species, the specimen samples were collected for herbarium preparation and identification was done in the laboratory by using 'Flora of the Presidency of Madras' [30]. For grass species the per cent cover per unit area was calculated through visual estimation rather than counting individual species. The percentage of invasive species *Lantana camara* cover per plot was also recorded through visual estimation at the time of study period. Visual estimation is fast, requires no specialized equipment, and can be adapted to plants of various growth forms [28, 29]. In addition to this, the number of cut stems and cowpats was recorded in the transects to assess the intensity of fuel wood collection and cattle grazing respectively in the study area.

Plant community variables such as species richness, Shannon's diversity *H'* and evenness *J* was calculated for the corridor landscape. Simple linear regression models were developed to test the influence of *Lantana camara,* fuelwood collection and cattle grazing on native plant diversity. In addition to this, species Importance Value Index (IVI) was calculated to identify the dominant species of the study area for both tree and non-tree classes.

For trees the IVI was calculated by using the formula; ***IVI of sp. i = relative density of sp. i + relative frequency of sp. i + relative dominance of sp. i.*** However, since data on relative dominance which is derived from basal area is not possible for non-trees, the IVI for undergrowth (non-trees) was calculated using the formula modified as ***IVI of sp. i = relative density of sp. i + relative frequency of sp. i.***

Local community considerations were also considered in addition to scientific data in prioritizing suitable native plant species for eco-restoration. This is because people from the landscape, especially *Soliga* tribals, possess sophisticated knowledge about biodiversity and traditional forest resource management practices [25, 31, 32]. Therefore, a participatory approach was employed to prioritize native plant species. Three Focus Group Discussions (FGD) were conducted in three corridor landscape dependent villages. In addition, a couple of informal interviews were also conducted. Questions were asked regarding corridors, wildlife, eco-restoration and presence of suitable plant species in the landscape.

3. Results

3.1. Plant community structure

Species richness and Shannon's diversity *H'* is relatively higher in tree class compared to shrub and herbaceous class. The evenness *J* is more or less similar between shrub and herbaceous class but relatively higher than tree class (**Table 1**). The corridor landscape had 92 tree species (belonging to 39 families), 75 shrub species (belonging to 41 families) and 185 species (belonging to 65 families). About 73.9% stems belong to different shrub species and 26.1% are saplings of woody species. In terms of total herbaceous stems enumerated in the study area, around 77.8% are herbaceous plants and 22.2% are woody seedlings.

3.1.1. Resource plants

The study area is endowed with rich plant resources. Out of 92 tree species, 10 species turned out to be important Non-timber forest products (NTFP) resource plants. They represented 2.5% of the total stems enumerated in the area. Among the NTFP category, fruits of *Phyllanthus indofischeri* ranked high. Nine tree species provided fuelwood (per. Interviews with local people) – and represented 13.5% of the total stems enumerated. Thirteen species were identified as important food resource for elephants (as mentioned in Refs. [33–35]), which represent 18% of total stems recorded from the study area (**Table 2**).

3.2. Species importance value or IVI

The study site was evaluated for importance value index of each species. For tree species, the top ten most common species found in the sampled area were *Anogeissus latifolia, Chloroxylon swietenia, Erythroxylon monogynum, Dalbergia lanceolaria, Strychnos potatorum, Naringi crenulata, Acacia chundra, Diospyros montana, Canthium travencoricum* and *Ixora arborea* (**Table 3**). Among 92 species, these 10 species contribute 52% of the total IVI (Appendix A).

For non-tree forms such as shrubs/saplings, the top ten and most common species found in the corridor landscape were *Lantana camara, Pterolobium hexapetalum, Dodonaea viscosa, Randia dumetorum, Chloroxylon swietenia, Erythroxylon monogynum, Zizyphus oenoplia, Fluggea leucopyrus, Eupatorium odoratum, Dolichandrone falcata* and *Pavetta indica* (**Table 4**). Among 75 species, these 10 species contribute 70% of the total IVI, of which *Lantana camara* alone contributes 32% (Appendix B).

For the seedlings/herbaceous plant group, the top ten most important species found in the corridor landscape were *Leucas martinicensis, Oxalis corniculata, Eupatorium odoratum, Lantana*

Community variable	Tree (mean ± se) Per 0.1 ha (n = 64)	Shrub (mean ± se) Per 10 m² (n = 128)	Herb (mean ± se) Per m² (n = 512)	Grass cover (mean ± se) percent/m² (n = 512)
Species richness	12.48 ± 0.53	6.13 ± 0.28	8.52 ± 0.14	–
Shannon's *H'*	2.06 ± 0.05	1.39 ± 0.05	1.72 ± 0.02	–
Evenness *J*	0.69 ± 0.01	0.78 ± 0.0	0.74 ± 0.006	–
Density	42.76 ± 3.36	21.15 ± 1.32	37.89 ± 1.05	44.90 ± 1.35

Table 1. Plant community variables among life forms (trees, shrubs, and herbs) of native vegetation in the corridor area.

http://dx.doi.org/10.5772/intechopen.72706

Sl. no.	Scientific name	Family	Importance
1	*Acacia chundra*	Mimosaceae	Fuelwood tree
2	*Anogeissus latifolia*	Combretaceae	Fuelwood tree
3	*Canthium travancoricum*	Rubiaceae	Fuelwood tree
4	*Chloroxylon swietenia*	Rutaceae	Fuelwood tree
5	*Erythroxylon monogynum*	Erythroxylaceae	Fuelwood tree
6	*Grewia asiatica*	Tiliaceae	Fuelwood tree
7	*Ixora arborea*	Rubiaceae	Fuelwood tree
8	*Randia dumetorum*	Rubiaceae	Fuelwood tree
9	*Ziziphus xylopyrus*	Rhamnaceae	Fuelwood tree
1	*Acacia sinuata*	Mimosaceae	NTFP plant (fruit)
2	*Azadirachta india*	Meliaceae	NTFP plant (fruit)
3	*Bombax ceiba*	Bombacaceae	NTFP (undeveloped fruit)
4	*Decalepis hamiltonii*	Asclepiadaceae	NTFP plant (root)
5	*Phoenix loureirii*	Arecaceae	NTFP plant (leaves)
6	*Phyllanthus indofischeri*	Euphorbiaceae	NTFP plant (fruit)
7	*Syzygium cumini*	Myrtaceae	NTFP plant (fruit)
8	*Tamarindus indica*	Fabaceae	NTFP plant (fruit)
9	*Terminalia bellerica*	Combretaceae	NTFP plant (fruit)
10	*Terminalia chebula*	Combretaceae	NTFP plant (fruit)
1	*Acacia chundra*	Fabaceae	Elephant food plant
2	*Acacia leucophlea*	Mimosaceae	Elephant food plant
3	*Acacia sinuata*	Mimosaceae	Elephant food plant
4	*Albizia amara*	Fabaceae	Elephant food plant
5	*Atylosia lineata*	Fabaceae	Elephant food plant
6	*Bambusa arundinacea*	Poaceae	Elephant food plant
7	*Capparis seperaria*	Capparaceae	Elephant food plant
8	*Commiphora caudata*	Burseraceae	Elephant food plant
9	*Dendrocalamas strictus*	Poaceae	Elephant food plant
10	*Grewia tilifolia*	Malvaceae	Elephant food plant
11	*Hardwickia binata*	Fabaceae	Elephant food plant
12	*Tectona grandis*	Verbenaceae	Elephant food plant
13	*Ziziphus xylopyrus*	Rhamnaceae	Elephant food plant

Table 2. List of fuelwood, NTFP, and elephant food plant species in the corridor area.

Dominant tree species	IVI value
Chloroxylon swietenia	32.89
Anogeissus latifolia	30.72
Erythroxylon monogynum	28.76
Acacia chundra	11.88
Dalbergia lanceolaria	11.48
Strychnos potatorum	10.56
Naringi crenulata	08.57
Diospyros montana	08.34
Ixora arborea	07.74
Canthium travancoricum	07.70

Table 3. Importance Value Index (IVI) for top ten tree species in the corridor landscape of BRT Tiger Reserve.

Non-tree forms	Dominant species	IVI value
Saplings/shrubs	*Lantana camara*	64.60
	Pterolobium hexapetalum	13.20
	Dodonia viscosa	11.92
	Randia dumetorum	09.68
	Chloroxylon swietenia	09.54
	Erythroxylon monogynum	07.63
	Ziziphus oenoplia	07.52
	Fluggea leucopyrus	05.88
	Eupatorium odoratum	05.65
	Dolichandrone falcata	05.47
Seedlings/herbs	*Leucas martinicensis*	16.81
	Oxalis corniculata	12.40
	Eupatorium odoratum	11.00
	Lantana camara	10.96
	Evolvulus alsinoides	05.68
	Atylosia lineata	04.59
	Randia dumetorum	04.57
	Justicia simplex	04.10
	Crotalaria calycina	03.98
	Ziziphus oenoplia	03.10

Table 4. Importance Value Index (IVI) for top ten non-tree species in the corridor landscape of BRT Tiger Reserve.

camara, Evolvulus alsinoides, Atylosia lineata, Randia dumetorum, Justicia simplex, Crotalaria calycina and *Ziziphus oenoplia* (**Table 4**). Among 185 species, these 10 species contribute 38% of the total IVI (Appendix C).

The problematic invasive weeds of the landscape, such as *Lantana camara* and *Eupatoruim odoratum* are contributing significantly toward total IVI in both shrubs and herbs categories. *Lantana camara* contributes 32.30% and 5.47% for total IVI of shrubs and herbs respectively, whereas *Eupatoruim odoratum* contributes 2.82% and 5.89% for total IVI of shrubs and herbs respectively. This indicates the extent of invasion of weeds in the landscape.

3.3. Relationship between vegetation diversity and habitat characteristics

The data was analyzed for relationships between one of the community variables such as vegetation diversity - of trees, shrubs and herbs - (as a response variable) with three habitat covariates

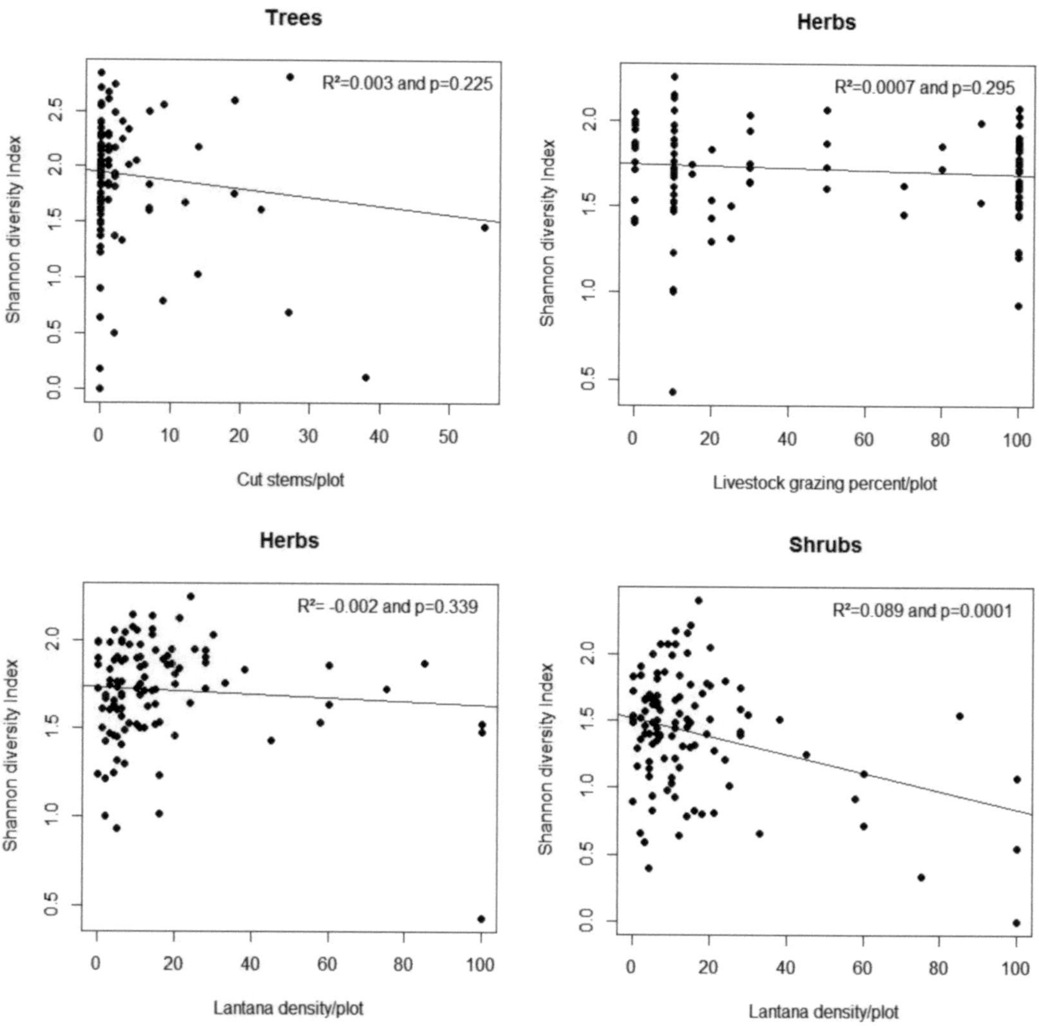

Figure 3. Relationships between species diversity (H') and three habitat characteristics (fuelwood collection, livestock grazing and *Lantana camara* density). Cut stems/plot implies fuelwood collection in the landscape.

such as fuelwood collection, livestock grazing intensity and invasive species – *Lantana camara* density (as predictor variables). The (four) models were developed to test the relationship between Diversity (*H'*) of- (i) trees vs. fuelwood collection, (ii) shrubs vs. *Lantana camara* density, (iii) herbs vs. *Lantana camara* density and (iv) herbs vs. grazing intensity of livestock.

Even though no statistically significant linear dependence of the mean of **y** on **x** was detected (the p-values are >0.05 for all relationships except for Shannon's diversity vs. *Lantana camara* density in shrubs) the slope (regression coefficients) shows a negative trend (**Figure 3**). The negative (marked in minus symbol) slope coefficient value for (i) trees vs. fuelwood collection is −0.007, (ii) shrubs vs. *Lantana camara* density is −0.006, (iii) herbs vs. *Lantana camara* density is −0.001 and (iv) herbs vs. grazing intensity of livestock −0.005. This indicates that fuelwood collection, cattle grazing and the density of invasive species like *Lantana camara* affects the species diversity (H') of life forms (trees, shrubs and herbaceous species) in the corridor landscape.

4. Discussion

Species richness is often treated not only as a measure of biodiversity [36] but also quality of the ecosystem and recovery of forest from disturbances such as logging [37–39]. The corridor is in the dry deciduous and scrub forest harboring 92 tree species in the sampled area, representing approximately 12% of plant species of the entire BRT forest enumerated [40]. The study site had around 10 NTFP species that provide partial household income for people in the corridor landscape; 12% for *Soligas* and 7% for non-*Soligas* [27]. The fruit of Indian Gooseberry tree is not only serves as a livelihood source for local people but also as an important dietary component for wild animals during the lean season [41–43]. As a result around 17% of *amla* sapling stems are re-sprouts in the study area. As in Ref. [44], fire and grazing in BRT could be the drivers of the high proportion of re-sprout as part of the demography.

The study result shows that vegetation diversity decreased with increase in fuelwood collection (in tree class), livestock grazing and invasive species (in non-tree class). Subsequently it will severely affect not only the plant community structure and regeneration [45, 46] but also habitat quality of the landscape [24], genetic structure of NTFPs at population level [47] and increment of woody vegetation [48]. *Lantana camara* is affecting native vegetation mainly of herbaceous class and shrub species, and is responsible for significant reduction in species richness and diversity [49]. As in Ref. [50] the study result from BRT forest showed that *Lantana camara* is the major driver impacting the demographic pattern of species such as *P. emblica* and *P. indofischeri*. This could be due to poor survival of light demanding seedlings of native tropical dry forest species under the conditions of high *Lantana camara* abundance and shade [51]. If the present scenario continues for a long period of time, it will gradually reduce forest regeneration rates and thus lead to impaired sustainability of the corridors [49, 52, 53].

4.1. Prioritized plant species for eco-restoration: a socio-ecological approach

Globally, conceptual models for restoration of biodiversity have highlighted the importance of regional plant source pool and framework species in restoration [54–56]. Regional plant species are more important for eco-restoration, because the type of forest occurring naturally in a place is the result of climatic, edaphic, topographic, and biotic factors of the locality [22, 23].

Out of 92 tree species, 10 species contribute 52% of the total IVI of the corridor landscape. Among the 10 species *Anogeissus latifolia, Canthium travancoricum, Erythroxylon monogynum* and *Ixora arborea* are the top five species which have been exploited for fuelwood. People prefer these trees as firewood due to their calorific value, ease of carrying as headload, and frequency of availability. Though species such as *Cassia spectabilis* and *Eucalyptus* sp. could form good fuelwood and timber trees respectively they are not collected by people as they are planted by the Forest Department. Some of the other tree species with high IVI in this landscape are not preferred either as fuelwood species or as domestic timber requirements due to multiple reasons. For instance, *Chloroxylon swietenia, Acacia chundra,* and *Strychnos potatorum* are tree species with thick/rough bark and are uncomfortable to carry as headload. Similarly *Diospyros montana* is not harvested for fuelwood because of the belief that doing so could splinter the family by inciting fights between family members. Similarly, people believe that *Terminalia bellerica* is one of the sacred trees in the landscape and belongs to the god *Shani Devaru*, (a local deity regarded as an incarnation of *Shiva*). Hence, we have shortlisted *Anogeissus latifolia* as a dominant and firewood tree species, and *Terminalia crenulata, Dalbergia lanceolaria* and *Albizia odoratissima* as timber tree species for vegetation enrichment. Since *Phyllanthus indofischeri* and *Terminalia bellerica* are major NTFP species that serve as a source of livelihood for local people [41] and also form part of the dietary requirement for ungulates during the lean season, people generally do not cut these trees for fuelwood. So, we have shortlisted these two species also for vegetation enrichment. Since honey is a major NTFP in this landscape, people suggested the planting of one nectar yielding tree species for honey bees in the landscape such as *Pterocarpus marsupium*. In addition to these, *Acacia chundra, Hardwickia binata* and *Bambusa arundinacea* were identified and shortlisted as important plant sources of elephant's food in the landscape [33–35].

Ten suitable native plant species were identified for vegetation enrichment based on their Important Value Index, ecological importance and recommendation by the community. Our research prioritized similar plant species for restoration such as *Anogeissus latifolia* (dominant tree and source of firewood), *Terminalia crenulata, Dalbergia lanceolaria* and *Albizia odoratissima* (timber trees), *Phyllanthus indofischeri* and *Terminalia bellirica* (NTFP trees), *Pterocarpus marsupium* (nectar source for honey bees), *Acacia chundra, Hardwickia binata* and *Bambusa arundinacea* (elephant food plants).

4.2. Species selected for clonal propagation

The plant species such as *Bambusa aurindinacea, Tectona grandis, Gmelina arborea* and *Dalbergia sissoo* in the corridor landscape may have the capability to propagate through clonal methods. Clonally propagated species (CPS) have the capacity to tolerate adverse conditions and give significantly better growth rates, and better disease resistance with most desirable timber traits [57]. In addition to this, clonal propagation trait not only could persist and maintain species richness but also retain genetic diversity of the species in the forests even after experiencing disturbance in the form of forest fire, grazing, and harvesting pressure from fuelwood collection [58, 59]. Since clonal propagation of dry tropical forest trees influence the tree species composition and demography, we suggested planting CPS, including bamboo along the forest boundary and teak in the farmland of the study area.

4.3. Nursing plants

Most of the forest landscapes in BRT have been subjected to different kinds of forest management practices such as shifting cultivation, logging, monoculture plantation, etc., both by the indigenous community and the State Forest Department in the past. This makes it more complex when it comes to understanding the structure, composition and successional status of native species [24, 25]. However, in eco-restoration, in order to improve the performance of target species, the "nursing" procedure seems to be promising, and shows enhanced plant survival and growth [18]. Therefore, in the same landscape, two native species, *Pterolobium hexapetalum* and *Dodonaea viscosa* were identified. These could play the role of nursing plants as they cover the native shrub and sapling communities extensively in more open forested areas. Being a prickly straggler, *Pterolobium hexapetalum* is not grazed by cattle and other ungulates. Likewise, *Dodonaea viscosa*, a bushy plant, is a pioneer species that is not eaten by cattle or other ungulates. Based on our field observations, we believe that these two native plants *P. hexapetalum* and *D. viscosa* could play the role of nursing by protecting seedlings from grazing and browsing, and influence the regeneration of tree seedlings and saplings.

5. Conclusion

In a human-dominated forest landscape like BRT, corridors have been subjected to severe anthropogenic disturbances and poor management. Fuelwood collection and livestock grazing coupled with invasive species *Lantana camara* have affected the vegetation dynamics of the corridor landscape. This will indirectly affect not only the dependent animal community but also the livelihoods of local people at some point in the same landscape. Our study has provided base line information on composition and size of the regional plant species pool, and also selected 10 native plant species for vegetation enrichment as part of eco-restoration in the corridor. Active and large scale *Lantana camara* removal coupled with enrichment planting activity needs to be initiated in and around the corridors to improve the habitat quality of the corridor landscape. Exploring the possibilities of using native shrub plants such as *Pterolobium hexapetalum* and *Dodonaea viscosa* as nursing plants to promote the survival rate of saplings of tree species could be one of the strategies. Convergence in the form of collaboration with local community, local institutions, local stakeholders, civil society, government and non-government research organizations is essential for improved protection and sustainable management of these important corridors. Such collaboration may help to increase the likelihood of persistence of animal populations by providing functional connectivity between the fragments. In fact the local community showed interest in establishing decentralized nurseries in the landscape to raise the selected plant species on incentive basis in collaboration with the Forest Department and the Village Panchayat. At a time of unprecedent habitat destruction, this could promote not only local participation and co-management of the wildlife corridor in a human-dominated forest landscape but also contribute toward 'UN-REDD Programme Strategic Framework' which is aiming to enhance carbon stocks in degraded forests [60].

http://dx.doi.org/10.5772/intechopen.72706

Acknowledgements

We acknowledge financial support received from USAID India (AID-386 A-14-00011) and research permission from the Karnataka State Forest Department to carry out field work.

Appendices

Appendix A. IVI of tree species in the corridor landscape of Biligiri Rangaswamy Temple Tiger Reserve

Sl. no.	Botanical name	Family	Relative density	Relative frequency	Relative basal area	IVI
1	*Chloroxylon swietenia*	Rutaceae	17.21	6.13	9.55	32.89
2	*Anogeissus latifolia*	Combretaceae	13.08	5.63	12.01	30.72
3	*Erythroxylon monogynum*	Erythroxylaceae	14.58	6.26	7.93	28.76
4	*Acacia chundra*	Mimosaceae	4.86	4.38	2.65	11.89
5	*Dalbergia lanceolaria*	Fabaceae	2.74	3.63	5.11	11.48
6	*Strychnos potatorum*	Strychnaceae	4.02	3.38	3.17	10.57
7	*Naringi crenulata*	Rutaceae	3.22	3.13	2.23	8.57
8	*Diospyros montana*	Ebenaceae	2.45	4.13	1.77	8.35
9	*Ixora arborea*	Rubiaceae	2.67	3.75	1.32	7.74
10	*Canthium travancoricum*	Rubiaceae	2.92	3.00	1.78	7.70
11	*Randia dumetorum*	Rubiaceae	2.52	3.25	0.91	6.69
12	*Dalbergia latifolia*	Fabaceae	0.47	0.75	5.25	6.47
13	*Atlantia monophylla*	Rutaceae	2.67	2.63	0.92	6.21
14	*Acacia leucophlea*	Mimosaceae	0.62	0.63	4.35	5.60
15	*Lantana camara*	Verbenaceae	0.66	2.25	2.39	5.30
16	*Diospyros melanoxylon*	Ebenaceae	1.57	2.63	0.50	4.70
17	*Ziziphus oenoplia*	Rhamnaceae	1.94	2.38	0.38	4.70
18	*Prosopis cineraria*	Fabaceae	0.44	1.00	2.64	4.08
19	*Albizia amara*	Fabaceae	0.69	1.50	1.84	4.04
20	*Stereospermum personatum*	Bignoniaceae	0.55	1.50	1.42	3.47
21	*Cassine glauca*	Celastraceae	1.06	1.63	0.71	3.39
22	*Cassia fistula*	Caesalpinaceae	0.91	2.13	0.31	3.36
23	*Premna tometosa*	Verbenaceae	0.69	1.63	0.97	3.30
24	*Phyllanthus indofischeri*	Euphorbiaceae	0.69	1.63	0.95	3.28
25	*Bambusa arundinacea*	Poaceae	1.24	0.63	1.33	3.20
26	*Grewia tiliifolia*	Tiliaceae	0.99	1.75	0.39	3.12

Sl. no.	Botanical name	Family	Relative density	Relative frequency	Relative basal area	IVI
27	*Ferronia elephantum*	Rutaceae	0.44	1.25	1.24	2.93
28	*Bauhinia purpurea*	Fabaceae	0.15	0.25	2.51	2.91
29	*Albizia odoratissima*	Fabaceae	0.15	0.50	2.21	2.86
30	*Vitex altissima*	Verbenaceae	0.55	1.38	0.88	2.80
31	*Diospyros* sp.	Ebenaceae	0.11	0.25	2.20	2.56
32	*Morinda tinctoria*	Rubiaceae	0.62	1.13	0.78	2.53
33	*Canthium parviflorum*	Rubiaceae	0.80	1.50	0.21	2.51
34	*Maytenus emarginata*	Celastraceae	0.80	1.38	0.29	2.47
35	*Ziziphus xylopyrus*	Rhamnaceae	0.84	1.25	0.36	2.45
36	*Dolichandrone falcata*	Bignoniaceae	1.06	0.88	0.15	2.08
37	*Gmelina arborea*	Verbenaceae	0.15	0.50	1.13	1.78
38	*Aglaia odoratissima*	Meliaceae	0.37	0.88	0.45	1.69
39	*Dodonaea viscosa*	Sapindaceae	0.69	0.88	0.10	1.67
40	*Commiphora caudata*	Burseraceae	0.26	0.63	0.77	1.65
41	*Hardwickia binata*	Caesalpinaceae	0.29	1.00	0.24	1.54
42	*Pterocarpus marsupium*	Fabaceae	0.18	0.63	0.70	1.51
43	*Schleichera oleosa*	Sapindaceae	0.04	0.13	1.30	1.47
44	*Unid2*	Unid	0.07	0.25	1.12	1.45
45	*Garuga pinnata*	Meliaceae	0.11	0.25	1.06	1.42
46	*Terminalia paniculata*	Combretaceae	0.33	0.63	0.42	1.37
47	*Celtis tetrandra*	Ulmaceae	0.11	0.25	0.99	1.35
48	*Haldina cordifolia*	Rubiaceae	0.11	0.38	0.81	1.30
49	*Acacia sinuata*	Mimosaceae	0.77	0.38	0.12	1.26
50	*Flacourtia montana*	Flacourtiaceae	0.22	0.25	0.63	1.10
51	*Terminalia bellirica*	Combretaceae	0.07	0.25	0.71	1.03
52	*Ficus* sp.	Moraceae	0.22	0.25	0.55	1.02
53	*Terminalia chebula*	Combretaceae	0.26	0.63	0.10	0.99
54	*Gmelina asiatica*	Verbenaceae	0.29	0.63	0.06	0.97
55	*Boswellia serrata*	Burseraceae	0.29	0.38	0.27	0.94
56	*Pterolobium hexapetalum*	Caesalpinaceae	0.26	0.63	0.05	0.93
57	*Caralluma umbellata*	Asclepiadaceae	0.37	0.38	0.18	0.92
58	*Azadirachta india*	Meliaceae	0.22	0.50	0.13	0.85
59	*Capparis seperaria*	Capparaceae	0.18	0.50	0.15	0.84
60	*Acacia nilotica*	Fabaceae	0.18	0.13	0.53	0.84

Sl. no.	Botanical name	Family	Relative density	Relative frequency	Relative basal area	IVI
61	*Ziziphus jujuba*	Rhamnaceae	0.18	0.25	0.36	0.79
62	*Cadaba fruticosa*	Capparaceae	0.11	0.38	0.28	0.76
63	*Santalum album*	Santalaceae	0.22	0.50	0.03	0.75
64	*Spondias pinnata*	Anacardiaceae	0.04	0.13	0.59	0.75
65	*Holarrhena antidycenterica*	Apocynaceae	0.15	0.50	0.06	0.71
66	*Butea monosperma*	Fabaceae	0.11	0.38	0.21	0.70
67	*Pongamia pinnata*	Fabaceae	0.11	0.25	0.31	0.67
68	*Acacia* sp.	Mimosaceae	0.07	0.25	0.34	0.66
69	*Dendrocalamas* sp.	Poaceae	0.26	0.38	0.03	0.66
70	*Flacourtia indica*	Flacourtiaceae	0.22	0.38	0.03	0.63
71	*Gardenia gammifera*	Rubiaceae	0.11	0.38	0.12	0.60
72	*Anacardium occidentale*	Anacardiaceae	0.04	0.13	0.42	0.58
73	*Strychnos* sp.	Strychnaceae	0.26	0.25	0.07	0.58
74	*Cleistanthus* sp.	Phyllanthaceae	0.11	0.38	0.08	0.57
75	*Wrightia tinctoria*	Apocynaceae	0.11	0.38	0.07	0.55
76	*Bridelia retusa*	Euphorbiaceae	0.11	0.38	0.06	0.55
77	*Terminalia crenulata*	Combretaceae	0.07	0.13	0.22	0.41
78	*Memecylon umbellatum*	Melastomataceae	0.15	0.25	0.01	0.41
79	*Bombax cieba*	Bombacaceae	0.04	0.13	0.19	0.35
80	*Tamarindus indiaca*	Fabaceae	0.04	0.13	0.18	0.34
81	*Carissa carandas*	Apocynaceae	0.07	0.25	0.02	0.34
82	*Celastrus paniculata*	Celastraceae	0.07	0.25	0.01	0.33
83	*Unid3*	Unid	0.04	0.13	0.03	0.20
84	*Erythrina variegata*	Fabaceae	0.04	0.13	0.03	0.19
85	*Unid1*	Unid	0.04	0.13	0.01	0.18
86	*Mallotus philippensis*	Euphorbiaceae	0.04	0.13	0.01	0.18
87	*Lagerstromia parviflora*	Lythraceae	0.04	0.13	0.01	0.17
88	*Grewia asiatica*	Tiliaceae	0.04	0.13	0.01	0.17
89	*Pyrenacantha volubilus*	Icacinaceae	0.04	0.13	0.01	0.17
90	*Chionanthus malabaricus*	Olacaceae	0.04	0.13	0.00	0.17
91	*Cocculus* sp.	Menispermaceae	0.04	0.13	0.00	0.17
92	*Syzygium cuminii*	Myrtaceae	0.04	0.13	0.00	0.16
			100	**100**	**100**	**300**

Appendix B. IVI of shrub species (includes saplings of woody plants) in the corridor landscape of Biligiri Rangaswamy Temple Tiger Reserve. 'Relative basal area' will not be considered for non-tree species

Sl. no.	Botanical name	Family	Relative density	Relative frequency	IVI
1	*Lantana camara*	Verbenaceae	51.60	13.00	64.60
2	*Pterolobium hexapetalum*	Caesalpinaceae	5.87	7.33	13.20
3	*Dodonia viscosa*	Sapindaceae	6.58	5.33	11.92
4	*Randia dumetorum*	Rubiaceae	3.90	5.78	9.68
5	*Chloroxylon swietenia*	Rutaceae	3.09	6.44	9.54
6	*Erythroxylon monogynum*	Erythroxylaceae	2.07	5.56	7.63
7	*Ziziphus oenoplia*	Rhamnaceae	2.18	5.33	7.52
8	*Fluggea leucopyrus*	Phyllanthaceae	1.77	4.11	5.88
9	*Eupatorium odoratum*	Asteraceae	4.20	1.44	5.65
10	*Dolichandrone falcata*	Bignoniaceae	2.36	3.11	5.47
11	*Pavetta indica*	Rubiaceae	2.33	2.78	5.10
12	*Toddalia asiatica*	Rutaceae	1.36	3.00	4.36
13	*Atlantia monophylla*	Rutaceae	1.29	2.33	3.62
14	*Acacia sinuata*	Mimosaceae	1.32	2.22	3.55
15	*Naringi crenulata*	Rutaceae	1.32	2.11	3.43
16	*Diospyros montana*	Ebenaceae	0.59	2.67	3.26
17	*Canthium travancoricum*	Rubiaceae	0.73	2.00	2.73
18	*Anogeissus latifolia*	Combretaceae	0.70	1.56	2.25
19	*Bambusa arundinacea*	Poaceae	0.45	1.33	1.78
20	*Ixora arborea*	Rubiaceae	0.45	1.22	1.67
21	*Flacourtia montana*	Flacourtiaceae	0.50	1.11	1.61
22	*Acacia chundra*	Mimosaceae	0.36	1.22	1.58
23	*Strychnos potatorum*	Strychnaceae	0.39	1.11	1.50
24	*Cassia fistula*	Caesalpinaceae	0.25	1.22	1.47
25	*Albizia amara*	Fabaceae	0.27	1.11	1.38
26	*Grewia tiliifolia*	Tiliaceae	0.23	1.00	1.23
27	*Santalum album*	Santalaceae	0.21	0.89	1.10
28	*Capparis seperaria*	Capparaceae	0.25	0.78	1.03
29	*Wrightia tinctoria*	Apocynaceae	0.20	0.78	0.97
30	*Grewia asiatica*	Tiliaceae	0.25	0.67	0.92

Sl. no.	Botanical name	Family	Relative density	Relative frequency	IVI
31	*Canthium parviflorum*	Rubiaceae	0.23	0.67	0.90
32	*Diospyros melanoxylon*	Ebenaceae	0.16	0.67	0.83
33	*Jasminum roxberghianum*	Oleaceae	0.23	0.44	0.68
34	*Cipadessa baccifera*	Meliaceae	0.20	0.44	0.64
35	*Maytenus emarginata*	Celastraceae	0.13	0.44	0.57
36	*Dalbergia lanceolaria*	Fabaceae	0.09	0.44	0.53
37	*Argyreia cuneata*	Convolvulaceae	0.07	0.44	0.52
38	*Memecylon umbellatum*	Melastomataceae	0.14	0.33	0.48
39	*Flacourtia indica*	Flacourtiaceae	0.13	0.33	0.46
40	*Ferronia elephantum*	Rutaceae	0.07	0.33	0.40
41	*Acacia leucophlea*	Mimosaceae	0.05	0.33	0.39
42	*Carissa carandas*	Apocynaceae	0.05	0.33	0.39
43	*Diospyros* sp.	Ebenaceae	0.05	0.33	0.39
44	*Premna tometosa*	Verbenaceae	0.05	0.33	0.39
45	*Stereospermum personatum*	Bignoniaceae	0.05	0.33	0.39
46	*Solanum torvum*	Solanaceae	0.09	0.22	0.31
47	*Azadirachta india*	Meliaceae	0.05	0.22	0.28
48	*Caralluma umbellata*	Asclepiadaceae	0.05	0.22	0.28
49	*Cassine glauca*	Celastraceae	0.05	0.22	0.28
50	*Maesa indica*	Myrsinaceae	0.05	0.22	0.28
51	*Prosopis cineraria*	Fabaceae	0.05	0.22	0.28
52	*Albizia odoratissima*	Fabaceae	0.04	0.22	0.26
53	*Celastrus paniculata*	Celastraceae	0.04	0.22	0.26
54	*Cycas* sp.	Cycadaceae	0.04	0.22	0.26
55	*Gardenia gammifera*	Rubiaceae	0.04	0.22	0.26
56	*Holarrhena antidycenterica*	Apocynaceae	0.04	0.22	0.26
57	*Jasminum* sp.	Oleaceae	0.04	0.22	0.26
58	*Opuntia elatior*	Cactaceae	0.04	0.22	0.26
59	*Phyllanthus emblica*	Euphorbiaceae	0.04	0.22	0.26
60	*Senna auriculata*	Fabaceae	0.04	0.22	0.26
61	*Tectona grandis*	Verbenaceae	0.04	0.22	0.26
62	*Vitex altissima*	Verbenaceae	0.04	0.22	0.26
63	*Barleria* sp.	Acanthaceae	0.14	0.11	0.25
64	*Phoenix loureirii*	Arecaceae	0.11	0.11	0.22
65	*Aglaia odoratissima*	Meliaceae	0.02	0.11	0.13

Sl. no.	Botanical name	Family	Relative density	Relative frequency	IVI
66	*Cocculus* sp.	Menispermaceae	0.02	0.11	0.13
67	*Decalepis hamiltonii*	Apocynaceae	0.02	0.11	0.13
68	*Dendrocalamas* sp.	Poaceae	0.02	0.11	0.13
69	*Givotia rottlerformis*	Euphorbiaceae	0.02	0.11	0.13
70	*Hardwickia binata*	Caesalpinaceae	0.02	0.11	0.13
71	*Jasminum angustifolium*	Oleaceae	0.02	0.11	0.13
72	*Lagerstromia parviflora*	Lythraceae	0.02	0.11	0.13
73	*Pyrenacantha volubilus*	Icacinaceae	0.02	0.11	0.13
74	*Ximenia americana*	Olacaceae	0.02	0.11	0.13
75	*Ziziphus xylopyrus*	Rhamnaceae	0.02	0.11	0.13
			100	**100**	**200**

Appendix C. IVI of herbaceous species (includes seedlings of woody plants) in the corridor landscape of Biligiri Rangaswamy Temple Tiger Reserve. 'Relative basal area' will not be considered for non-tree species

Sl. no.	Botanical name	Family	Relative density	Relative frequency	IVI
1	*Leucas martinicensis*	Lamiaceae	12.75	4.06	16.81
2	*Oxalis corniculata*	Oxalidaceae	8.41	3.99	12.40
3	*Eupatorium odoratum*	Asteraceae	6.96	4.03	11.00
4	*Lantana camara*	Verbenaceae	5.32	5.64	10.96
5	*Evolvulus alsinoides*	Convolvulaceae	3.15	2.52	5.68
6	*Atylosia* sp.	Fabaceae	2.05	2.54	4.59
7	*Randia dumetorum*	Rubiaceae	1.79	2.77	4.57
8	*Justicia simplex*	Acanthaceae	2.33	1.76	4.10
9	*Crotalaria calycina*	Fabaceae	2.14	1.83	3.98
10	*Ziziphus oenoplia*	Rhamnaceae	1.40	2.36	3.76
11	*Sida acuta*	Malvaceae	2.33	1.33	3.66
12	*Ipomoea* sp.	Convolvulaceae	1.46	2.06	3.52
13	*Phyllanthus amarus*	Euphorbiaceae	1.29	2.22	3.51
14	*Atylosia lineata*	Fabaceae	2.43	1.05	3.48
15	*Urena lobata*	Malvaceae	1.46	1.67	3.14
16	*Anogeissus latifolia*	Combretaceae	0.84	2.18	3.02

Sl. no.	Botanical name	Family	Relative density	Relative frequency	IVI
17	*Desmodiastrum racemosum*	Fabaceae	1.30	1.54	2.84
18	*Jasmium angustifolium*	Oleaceae	1.17	1.49	2.66
19	*Barleria prionitis*	Acanthaceae	1.39	1.24	2.63
20	*Fluggea leucopyrus*	Phyllanthaceae	0.75	1.83	2.59
21	*Pterolobium hexapetalum*	Caesalpinaceae	0.81	1.72	2.53
22	*Cynotis arachnoidea*	Commelinaceae	1.15	1.31	2.46
23	*Triumfetta rhomboidea*	Tiliaceae	1.27	1.08	2.35
24	*Achyranthes aspera*	Verbenaceae	1.23	1.08	2.30
25	*Curculigo orchioides*	Hypoxidaceae	0.76	1.49	2.25
26	*Grewia asiatica*	Tiliaceae	0.74	1.44	2.18
27	*Jasminum roxberghianum*	Oleaceae	0.93	1.17	2.10
28	*Acacia chundra*	Mimosaceae	0.61	1.47	2.08
29	*Rhynchosia viscosa*	Fabaceae	1.15	0.92	2.07
30	*Euphorbia hirta*	Euphorbiaceae	1.02	0.96	1.98
31	*Ocimum americanum*	Lamiaceae	0.96	0.96	1.93
32	*Hemedesmus indicus*	Apocynaceae	0.80	1.10	1.90
33	*Gymnema sylvestre*	Asclepiadaceae	0.97	0.87	1.84
34	*Leucas aspera*	Lamiaceae	1.21	0.60	1.80
35	*Dolichandrone falcata*	Bignoniaceae	0.69	1.08	1.77
36	*Dodonia viscosa*	Sapindaceae	0.56	1.19	1.75
37	*Anaphalis subdecurrense*	Asteraceae	0.58	1.10	1.68
38	*Scilla* sp.	Asparagaceae	0.60	1.08	1.68
39	*Galactia tenuiflora*	Fabaceae	0.86	0.80	1.66
40	*Chloroxylon swietenia*	Rutaceae	0.57	1.01	1.58
41	*Senna auriculata*	Fabaceae	0.79	0.71	1.50
42	*Abutilon* sp.	Malvaceae	0.74	0.76	1.49
43	*Diospyros montana*	Ebenaceae	0.43	1.03	1.46
44	*Indigofera* sp.	Fabaceae	0.99	0.46	1.45
45	*Acacia sinuata*	Mimosaceae	0.61	0.83	1.43
46	*Senna occidenatlis*	Fabaceae	0.75	0.66	1.41
47	*Orthosiphon rubicundus*	Lamiaceae	0.59	0.78	1.37
48	*Toddalia asiatica*	Rutaceae	0.41	0.94	1.35
49	*Ixora arborea*	Rubiaceae	0.41	0.94	1.35
50	*Crepis* sp.	Asteraceae	0.94	0.25	1.19
51	*Barleria buxifolia*	Acanthaceae	0.37	0.73	1.10

Sl. no.	Botanical name	Family	Relative density	Relative frequency	IVI
52	*Stachytarpheta india*	Verbenaceae	0.60	0.50	1.10
53	*Asparagas gonocladus*	Asparagaceae	0.27	0.78	1.05
54	*Stenosiphonium russelianium*	Acanthaceae	0.51	0.53	1.03
55	*Bidens* sp.	Asteraceae	0.43	0.60	1.03
56	*Cissampelos pareira*	Menispermaceae	0.34	0.66	1.00
57	*Ageratum conyzoides*	Asteraceae	0.71	0.25	0.96
58	*Cynotis* sp.	Commelinaceae	0.57	0.39	0.96
59	*Erythroxylon monogynum*	Erythroxylaceae	0.24	0.66	0.90
60	*Prosopis cineraria*	Fabaceae	0.31	0.55	0.86
61	*Pavetta indica*	Rubiaceae	0.25	0.60	0.84
62	*Andrographis serpyllifolia*	Acanthaceae	0.35	0.46	0.80
63	*Atlantia monophylla*	Rutaceae	0.29	0.50	0.79
64	*Dalbergia lanceolaria*	Fabaceae	0.27	0.50	0.78
65	*Hyptis suaveolens*	Lamiaceae	0.51	0.25	0.76
66	*Mimosa pudica*	Mimosaceae	0.41	0.34	0.76
67	*Sida rhombifolia*	Malvaceae	0.31	0.41	0.72
68	*Dalbergia latifolia*	Fabaceae	0.24	0.48	0.72
69	*Maytenus emarginata*	Celastraceae	0.26	0.46	0.72
70	*Senna* sp.	Fabaceae	0.39	0.25	0.64
71	*Pteridium* sp.	Dennstaedtiaceae	0.56	0.07	0.63
72	*Albizia amara*	Fabaceae	0.16	0.46	0.62
73	*Bidens barbidens*	Asteraceae	0.27	0.30	0.57
74	*Indigofera tinctoria*	Fabaceae	0.19	0.37	0.56
75	*Parthenium hysterophorus*	Asteraceae	0.22	0.30	0.52
76	*Canthium parviflorum*	Rubiaceae	0.15	0.34	0.50
77	*Artemisia pallens*	Asteraceae	0.26	0.23	0.49
78	*Albizia odoratissima*	Fabaceae	0.15	0.30	0.45
79	*Croton* sp.	Euphorbiaceae	0.10	0.34	0.44
80	*Leucas* sp.	Lamiaceae	0.23	0.21	0.44
81	*Cipadessa baccifera*	Meliaceae	0.18	0.25	0.43
82	*Eradale gida**	Fabaceae	0.24	0.18	0.43
83	*Mimosa* sp.	Mimosaceae	0.20	0.23	0.43
84	*Naringi crenulata*	Rutaceae	0.10	0.32	0.42
85	*Strobilanthes callosa*	Acanthaceae	0.27	0.11	0.39
86	*Malva* sp.	Malvaceae	0.20	0.18	0.38

http://dx.doi.org/10.5772/intechopen.72706

Sl. no.	Botanical name	Family	Relative density	Relative frequency	IVI
87	*Phyllanthus indofischeri*	Euphorbiaceae	0.12	0.25	0.38
88	*Solanum torvum*	Solanaceae	0.14	0.21	0.35
89	*Theriophonum* sp.	Araceae	0.15	0.18	0.33
90	*Cocculus* sp.	Menispermaceae	0.10	0.23	0.33
91	*Azima tetracantha*	Salvadoraceae	0.07	0.23	0.30
92	*Strychnos potatorum*	Strychnaceae	0.09	0.21	0.30
93	*Ocimum* sp.	Lamiaceae	0.25	0.05	0.30
94	*Stylosanthus* sp.	Fabaceae	0.11	0.18	0.29
95	*Pogostemon* sp.	Lamiaceae	0.08	0.21	0.29
96	*Abutilon hirtum*	Malvaceae	0.15	0.14	0.29
97	*Strychnos* sp.	Strychnaceae	0.06	0.23	0.29
98	*Cynanchum tunicatum*	Asclepiadaceae	0.08	0.21	0.28
99	*Jasminum* sp.	Oleaceae	0.13	0.14	0.27
100	*Pyrenacantha volubilus*	Icacinaceae	0.12	0.14	0.26
101	*Crotalaria* sp.	Fabaceae	0.11	0.14	0.25
102	*Ziziphus xylopyrus*	Rhamnaceae	0.06	0.18	0.25
103	*Santalum album*	Santalaceae	0.09	0.14	0.23
104	*Flacourtia montana*	Flacourtiaceae	0.07	0.16	0.23
105	*Lantana indica*	Verbenaceae	0.06	0.16	0.22
106	*Diospyros melanoxylon*	Ebenaceae	0.06	0.16	0.22
107	*Sida* sp.	Malvaceae	0.09	0.11	0.20
108	*Ferronia yesphantum*	Rutaceae	0.06	0.14	0.19
109	*Dioscorea oppositifolia*	Dioscoreaceae	0.05	0.14	0.19
110	*Sansevieria trifasciata*	Asparagaceae	0.07	0.11	0.19
111	*Ceropegia* sp.	Apocynaceae	0.06	0.11	0.17
112	*Thotti**	Unidentified	0.08	0.09	0.17
113	*Helicteres isora*	Malvaceae	0.04	0.11	0.15
114	*Pterocarpus marsupium*	Fabaceae	0.04	0.11	0.15
115	*Plectranthus amboinicus*	Lamiaceae	0.10	0.05	0.15
116	*Barleria* sp.	Acanthaceae	0.06	0.09	0.15
117	*Hardwickia binata*	Fabaceae	0.03	0.11	0.15
118	*Maesa indica*	Myrsinaceae	0.05	0.09	0.14
119	*Asaparagus racemosus*	Asparagaceae	0.03	0.11	0.14
120	*Mallotus philippensis*	Euphorbiaceae	0.03	0.11	0.14
121	*Stereospermum personatum*	Bignoniaceae	0.03	0.11	0.14

Sl. no.	Botanical name	Family	Relative density	Relative frequency	IVI
122	*Rauvolfia serpentina*	Apocynaceae	0.09	0.05	0.14
123	*Bambusa arundinacea*	Poaceae	0.05	0.09	0.14
124	*Ocimum tenuiflorum*	Lamiaceae	0.05	0.09	0.14
125	*Schleichera oleosa*	Sapindaceae	0.04	0.09	0.13
126	*Nela bhuthale**	Unidentified	0.08	0.05	0.13
127	*Cryptolepis buchnani*	Asclepiadaceae	0.04	0.09	0.13
128	*Memecylon umbellatum*	Melastomataceae	0.03	0.09	0.12
129	*Nicandra physalodes*	Solanaceae	0.05	0.07	0.12
130	*Padavara baale**	Unidentified	0.03	0.09	0.12
131	*Cassia fistula*	Caesalpinaceae	0.02	0.09	0.11
132	*Wrightia tinctoria*	Apocynaceae	0.02	0.09	0.11
133	*Celastrus paniculata*	Celastraceae	0.05	0.05	0.10
134	*Canthium travancoricum*	Rubiaceae	0.02	0.07	0.09
135	*Diospyros* sp.	Ebenaceae	0.02	0.07	0.09
136	*Argyreia cuneata*	Convolvulaceae	0.02	0.07	0.08
137	*Breynia retusa*	Euphorbiaceae	0.02	0.07	0.08
138	*Dioscorea* sp.	Dioscoreaceae	0.02	0.07	0.08
139	*Flacourtia indica*	Flacourtiaceae	0.02	0.07	0.08
140	*Gardenia gammifera*	Rubiaceae	0.02	0.07	0.08
141	*Actiniopteris radiata*	Pteridaceae	0.03	0.05	0.07
142	*Tephrosia* sp.	Fabaceae	0.03	0.05	0.07
143	*Vitex altissima*	Verbenaceae	0.03	0.05	0.07
144	*Caralluma umbellata*	Asclepiadaceae	0.02	0.05	0.07
145	*Cleistanthus* sp.	Phyllanthaceae	0.02	0.05	0.06
146	*Coccinia grandis*	Cucurbitaceae	0.02	0.05	0.06
147	*Elaeagnus conferta*	Elaeagnaceae	0.02	0.05	0.06
148	*Holarrhena antidycenterica*	Apocynaceae	0.02	0.05	0.06
149	*Phyllanthus virgatus*	Euphorbiaceae	0.02	0.05	0.06
150	*Acacia* sp.	Mimosaceae	0.01	0.05	0.06
151	*Argyreia cymosa*	Convolvulaceae	0.01	0.05	0.06
152	*Azadirachta india*	Meliaceae	0.01	0.05	0.06
153	*Millettia racemosa*	Fabaceae	0.01	0.05	0.06
154	*Odavara**	Unidentified	0.01	0.05	0.06

Sl. no.	Botanical name	Family	Relative density	Relative frequency	IVI
155	*Terminalia bellirica*	Combretaceae	0.01	0.05	0.06
156	*Terminalia crenulata*	Combretaceae	0.01	0.05	0.06
157	*Nada kappali**	Unidentified	0.03	0.02	0.05
158	*Carissa carandas*	apocynaceae	0.02	0.02	0.04
159	*Celtis tetrandra*	Ulmaceae	0.02	0.02	0.04
160	*Gmelina arborea*	Verbenaceae	0.02	0.02	0.04
161	*Acanthus* sp.	Acanthaceae	0.01	0.02	0.03
162	*Arda chandra**	Unidentified	0.01	0.02	0.03
163	*Eucalyptus globulus*	Myrtaceae	0.01	0.02	0.03
164	*Physalis minima*	Solanaceae	0.01	0.02	0.03
165	*Ximenia americana*	Olacaceae	0.01	0.02	0.03
166	*Antu huruligida**	Unidentified	0.01	0.02	0.03
167	*Antu pulle**	Unidentified	0.01	0.02	0.03
168	*Bombax cieba*	Bombacaceae	0.01	0.02	0.03
169	*Canthium* sp.	Rubiaceae	0.01	0.02	0.03
170	*Casearia tomentosa*	Salicaceae	0.01	0.02	0.03
171	*Cassine glauca*	Celastraceae	0.01	0.02	0.03
172	*Dendrocalamas* sp.	Poaceae	0.01	0.02	0.03
173	*Gloriosa superba*	Colchicaceae	0.01	0.02	0.03
174	*Hambu bhuthale**	Unidentified	0.01	0.02	0.03
175	*Hittina kudi**	Unidentified	0.01	0.02	0.03
176	*Huriyana hambu**	Unidentified	0.01	0.02	0.03
177	*Lamium* sp.	Lamiaceae	0.01	0.02	0.03
178	*Maathadakana hambu**	Unidentified	0.01	0.02	0.03
179	*Morinda tinctoria*	Rubiaceae	0.01	0.02	0.03
180	*Nela gorava**	Unidentified	0.01	0.02	0.03
181	*Premna tometosa*	Verbenaceae	0.01	0.02	0.03
182	*Sanna javana**	Lamiaceae	0.01	0.02	0.03
183	*Syzygium cuminii*	Myrtaceae	0.01	0.02	0.03
184	*Tectona grandis*	Verbenaceae	0.01	0.02	0.03
185	*Ziziphus jujuba*	Rhamnaceae	0.01	0.02	0.03
			100	**100**	**200**

Note: The botanical names of the * marked plant species were unidentified, instead the *Soliga* vernacular names were given.

Author details

Paramesha Mallegowda[1,2]*, Siddappa Setty[1] and Ganesan Rengaian[1]

*Address all correspondence to: paramesh@atree.org

1 Ashoka Trust for Research in Ecology and the Environment (ATREE), Bangalore, India

2 Manipal University, Udupi, India

References

[1] Lamb D. Large-scale ecological restoration of degraded tropical forest lands: The potential role of timber plantations. Restoration Ecology. 1998;**6**(3):271-279

[2] Brooks TM, Pimm SL, Oyugi JO. Time lag between deforestation and bird extinction in tropical forest fragments. Conservation Biology. 1999;**13**:1140-1150

[3] Dobson A, Lodge D, Alder J, Cumming GS, Keymer J, et al. Habitat loss, trophic collapse, and the decline of ecosystem services. Ecology. 2006;**87**(8):1915-1924

[4] Hilty JA, Lidicker WZ Jr, Merenlender AM. Corridor Ecology: The Science and Practice of Linking Landscapes for Biodiversity Conservation. USA: Island Press; 2006. 323 p

[5] WWF. How Effective are Protected Areas? A Report Prepared for the Seventh Conference of Parties of the Convention on Biological Diversity. Switzerland: WWF International; 2004. 24 p

[6] Menon V, Tiwari SK, Easa PS, Sukumar R. Right of Passage - Elephant Corridors of India. India: Wildlife Trust of India; 2005. 287 p

[7] MacKenzie CA, Chapman CA, Sengupta R. Spatial patterns of illegal resource extraction in Kibale National Park, Uganda. Environmental Conservation. 2011;**39**(1):38-50

[8] Hanski L. Metapopulation dynamics. Nature. 1998;**396**:41-49

[9] Aerts R, Honnay O. Forest restoration, biodiversity and ecosystem functioning. BMC Ecology. 2011;**11**:29

[10] Manjaribe C, Frasier CL, Rakouth B, Louis Jr. EE. Ecological Restoration and Reforestation of Fragmented Forests in Kianjavato, Madagascar. International Journal of Ecology. 2013;**2013**:1-12

[11] Lu Y, Ranjitkar S, Xu JC, Ou XK, Zhou YZ, et al. Propagation of native tree species to restore subtropical evergreen broad-leaved forests in SW China. Forests. 2016;**7**(1):12

[12] Restoring Forest Health and Habitat Connectivity for Wildlife [Internet]. 2016. Available from: http://www.conservationnw.org/what-we-do/columbiahighlands/forest-restoration [Accessed: 15-06-2016]

[13] National Mission for a Green India. Ministry of Environment and Forests, Government of India; 2010

[14] The Hindu [Internet]. 2010. Available from: http://www.thehindu.com/news/national/green-india-mission-to-double-afforestation-efforts-by-2020/article438506.ece [Accessed: 26-05-2010]

[15] The Times of India [Internet]. 2016. Available from: http://timesofindia.indiatimes.com/city/bengaluru/Govt-plans-to-buy-plantations-to-restore-elephant-corridors/articleshow/51103274.cms [Accessed: 23-02-2016]

[16] Rai SN. Nursery and Planting Techniques of Forest Trees in Tropical South-Asia. Dharwad, India: Punarvasu Publications; 1999

[17] Binkley CS. The Environmental Benefits of Forest Plantations. Research Note 2003-1. USA: GreenWood Resources; 2003. p. 10

[18] Padilla FM, Pugnaire FI. The role of nurse plants in the restoration of degraded environments. Frontiers in Ecology and the Environment. 2006;**4**(4):196-202

[19] Sayer J, Chokkalingam U, Poulsen J. The restoration of forest biodiversity and ecological values. Forest Ecology and Management. 2004;**201**:3-11

[20] Bremer LL, Farley KA. Does plantation forestry restore biodiversity or create green deserts? A synthesis of the effects of landuse transitions on plant species richness. Biodiversity and Conservation. 2010;**19**(14):3893-3915

[21] Pirard R, Seccoa LD, Warmanb R. Do timber plantations contribute to forest conservation? Environmental Science and Policy. 2016;**57**:22-130

[22] Shankar Raman TR, Mudappa D, Kapoor V. Restoring rainforest fragments: Survival of mixed-native species seedlings under contrasting site conditions in the Western Ghats, India. Restoration Ecology. 2009;**17**(1):137-147

[23] Khanna LS. Principles and Practice of Silviculture. India: Paperback Publishers; 2015

[24] Barve N, Kiran MC, Vanaraj G, Aravind NA, Rao D, Uma Shaanker R, Ganeshaiah KN, Poulsen JG. Measuring and mapping threats to a wildlife sanctuary in Southern India. Conservation Biology. 2005;**19**:122-130

[25] Mandal S, Rai ND, Madegowda C. Culture conservation and co-management: Strengthening Soliga stake in biodiversity conservation in Biligiri Rangaswamy Temple Wildlife Sanctuary, India. In: Verschuuren B, editor. Sacred Natural Sites: Conserving Nature and Culture. London: Earthscan and IUCN Publications Ltd; 2010. pp. 263-271

[26] Mallegowda P. Do wildlife corridors need protection? Sanctuary Asia. 2012;**32**:70-71

[27] Paramesha M. Functionality of wildlife corridors in the fragmented landscape of the Western Ghats, India: Implications for conservation and management [thesis]. India: Manipal University; 2015

[28] Kennedy KA, Addison PA. Some considerations for the use of visual estimates of plant cover in biomonitoring. Journal of Ecology. 1987;**75**:151-157

[29] Sutherland JW. Ecological Census Techniques, a hand book. UK: Cambridge University Press; 1996. 336 p

[30] Gamble JS. Flora of the Presidensy of Madras. London: Adlard and Son Ltd; 1935. 460 p

[31] Morab SG. The Soliga of Biligiri Rangana Hills. India: Anthropological Survey of India; 1977. 121 p

[32] Madegowda C. A study on life style of Soliga tribes at Biligiri Rangaswamy Temple Wildlife Sanctuary: A social work perspective [thesis]. India: University of Mysore; 2013

[33] Sukumar R. Ecology of the Asian elephant in southern India. II. Feeding habits and crop raiding patterns. Journal of Tropical Ecology. 1990;**6**:33-53

[34] Baskaran N, Balasubramanian M, Swaminathan S, Desai AA. Feeding ecology of the Asian Elephant *Elephas maximus* Linnaeus in the Nilgiri Biosphere Reserve, southern India. Journal of the Bombay Natural History Society. 2010;**107**(1):3-13

[35] Baskaran N, Desai AA. Frugivory and seed dispersal by the Asian Elephant *Elephas maximus* in the tropical forests of Nilgiri Biosphere Reserve, southern India. Journal of Threatened Taxa. 2013;**5**(14):4893-4897

[36] Magguran EA. Ecological Diversity and Its Measurement. Princeton: Princeton University Press; 1988. 179 p

[37] Connell JH. Diversity in tropical rain forests and coral reefs: High diversity of trees and corals is maintained only in a non-equilibrium state. Science. 1978;**199**:1302-1310

[38] Denslow JS. The effects of disturbance on tree diversity in tropical rain forest: The density effect. Ecological Applications. 1995;**5**:962-968

[39] Sheil D, Burslem D. Disturbing hypotheses in tropical forests. Trends in Ecology and Evolution. 2003;**18**:18-26

[40] Kammathy RV, Rao AS, Rao RS. A contribution towards a flora of Biligirirangan Hills. Mysore State Bulletin: Botanical Survey of India. 1967;**9**(1-4):206-234

[41] Setty RS. Ecology and productivity studies on some non-timber forest products of Biligiri Rangaswamy Wildlife Sanctuary [thesis]. India: University of Mysore; 2004

[42] John Singh AJT. Importance of fruit in the diet of chital in dry season. Journal of Bombay Natural History Society. 1981;**78**:594

[43] Prasad S, Chellam R, Krishnaswamy J, Goyal SP. Frugivory of *Phyllanthus emblica* at Rajaji National Park, North-West India. Current Science. 2004;**87**:1188-1190

[44] Ganesan R, Setty RS. Regeneration of amla, an important non-timber forest product from southern India. Conservation and Society. 2004;**2**:365-375

[45] Jha CS, Singh JS. Composition and dynamics of dry tropical forest in relation to soil texture. Journal of Vegetation Science. 1990;**1**:609-614

[46] Sagar R, Singh JS. Local plant species depletion in a tropical dry deciduous forest of northern India. Environmental Conservation. 2004;**31**(1):55-62

[47] Uma Shaanker R, Ganeshaiah KN, Nageswara Rao M, Aravind NA. Ecological consequences of forest use: From genes to ecosystem - A case study in the Biligiri Rangaswamy Temple Wildlife Sanctuary, South India. Conservation and Society. 2004;**2**:347-363

[48] Carmel Y, Kadmon R. Effects of grazing and topography on long-term vegetation changes in a Mediterranean ecosystem in Israel. Plant Ecology. 1999;**145**:243-254

[49] Sundaram B, Hiremath AJ. *Lantana camara* invasion in a heterogeneous landscape: Patterns of spread and correlation with changes in native vegetation. Biological Invasions. 2012;**14**(6):1127-1141

[50] Ticktin T, Ganesan R, Paramesha M, Setty S. Disentangling the effects of multiple anthropogenic drivers on the decline of two tropical dry forest trees. Journal of Applied Ecology. 2012;**49**:774-784

[51] Vieira DLM, Scariot A. Principles of natural regeneration of tropical dry forests for restoration. Restoration Ecology. 2006;**14**:11-20

[52] Spooner P, Lunt I, Robinson W. Is fencing enough? The short-term effects of stock exclusion in remnant grassy woodlands in southern NSW. Ecological Management and Restoration. 2002;**3**:117-126

[53] Kunwar RM, Sharma SP. Quantitative analysis of tree species in two community forests of Dolpa district, mid-west Nepal. Himalayan Journal of Sciences. 2004;**2**(3):23-28

[54] Zobel M, van der Maarel E, Dupré C. Species pool: The concept, its determination and significance for community restoration. Applied Vegetation Science. 1998;**1**:55-66

[55] Blakesley D, Hardwick K, Elliott S. Research needs for restoring tropical forests in Southeast Asia for wildlife conservation: Framework species selection and seed propagation. New Forests. 2002;**24**(3):165-174

[56] Brudvig LA. The restoration of biodiversity: Where has research been and where does it need to go? American Journal of Botany. 2011;**98**(3):549-558

[57] Lal P. Integrated development of farm -forestry plantations and wood based industries. Indian Forester. 2004;**130**(1):71-78

[58] Lopez RD, Davis CB, Fennessy MS. Ecological relationships between landscape change and plant guilds in depressional wetlands. Landscape Ecology. 2002;**17**:43-56

[59] Honnay O, Bossuyt B, Jacquemyn H, Hermy M. Forest fragmentation effects on patch occupancy and population viability of herbaceous plant species. New Phytologist. 2005;**166**:723-736

[60] UN-REDD Programme Strategic Framework 2016-20. Washington DC: UN-REDD Programme Fourteenth Policy Board Meeting; 2015. p. 45

Chapter 7

Apomixis in Neotropical Vegetation

Fabiana Firetti

Additional information is available at the end of the chapter

http://dx.doi.org/10.5772/intechopen.71856

Abstract

In neotropical ecosystems, genetic variability and diversity of plant species seem to be generated and maintained by sexual reproduction. However, apomixis appears as an alternative reproductive strategy in many plant families. Apomixis comprises the development of a new organism from an unreduced and unfertilized egg cell. Traditionally, it is considered a dead end of evolution due to the lack of genetic variability; however, processes such as *de novo* mutation, gene conversion, mitotic recombination and epigenetic drift may work as important sources of genetic variation during apomictic reproduction. Moreover, plant species show facultative apomixis, which allows the formation of sexual and asexual offspring. As a result, natural apomict populations show greater genotypic variability than expected from a clonal population. Also, asexual reproduction is considered one of the important attributes promoting diversity in angiosperms. Here, I review the occurrence of apomixis in several plant families that are well represented in neotropics and I infer the relation between apomixis and diversity of species. Many plant families that are common in the neotropical region show facultative apomixis associated with polyploidy and hybridization events. Apomixis seems to play a key role in the establishment of new evolutionary lineages in a wide variety of environmental conditions.

Keywords: asexual reproduction, diversity, genetic variability, plant diversity, tropical vegetation

1. Introduction

The tropics, subdivided into paleotropics and neotropics, include the region of the earth's surface between Tropic of Cancer (23°27′N) and the Tropic of Capricorn (23°27′S) and comprise about 40% of the earth's land surface [1]. As currently defined by Schultz [2], the neotropical ecozone extends from central Mexico in the north to southern Brazil in the south, including Central America, the Caribbean islands and most of South America. Comprising around 90,000–110,000 species of seed plants, the neotropics harbors about 37% of the world's species

[3]. This great diversity is allocated in different vegetation formations such as wet and dry forests, desert areas and savanna formations, which show very variable environmental conditions. A recent survey about world biodiversity has revealed that some countries of the neotropical region are among the world's 10 most megadiverse [4]. Brazil, the most megadiverse country in the world considering plants and fungi, presents its biodiversity distributed in six major phytogeographical domains. In addition, each phytogeographical domain presents a wide variety of vegetation types, habitats and microhabitats that are occupied by a wide diversity of plants, fungi and animals [4].

Many mechanisms have been proposed to have played a role in the historical assembly and evolution of neotropical biodiversity [3]. Among these mechanisms, the plant-animal interactions seem to play a relevant role in the generation of genetic diversity of plant species, through gene flow conducted via pollination and seed dispersal [3]. In addition, genetic variability seems to be generated and maintained also by the predominance of genetically controlled self-incompatibility systems. Self-incompatibility (SI) has been found to be the main breeding system in forest [5, 6], savanna [7] and semi-arid [8] communities. However, most of these studies evaluated the breeding systems of tree species. In herbaceous community and for species in disturbed areas usually with an inefficient pollination fauna, the lowest frequencies of self-incompatibility have been reported [9].

Despite the predominance of sexual reproduction, a growing number of studies on breeding systems have demonstrated the presence of apomixis as an alternative reproductive strategy of the plants of neotropical biomes [10]. Apomixis, the asexual reproduction via seed, comprises the development of a new organism from an unreduced and unfertilized egg cell [11]. It prevents meiosis and fertilization leading to a clonal progeny and presents a sexual development that is deregulated at several stages where meiosis is impaired or absent resulting in an unreduced female gametophyte, fertilization is avoided leading to the production of an autonomous embryo and the endosperm development is autonomous or sexual [12, 13]. Thus, three elements are commonly observed in apomicts: (1) the generation of a cell capable of forming an embryo without prior meiosis (apomeiosis); (2) the spontaneous, fertilization-independent development of the embryo (parthenogenesis); (3) and the capacity to either produce endosperm autonomously or to use a pseudogamous endosperm derived from fertilization [11, 13].

Studies suggest that apomixis may emerge from the sexual system deregulated by genetic and epigenetic modifications, which entails in the alteration or omission of stages of the sexual development [14–16]. The genomic changes can often be a response to the consequences of genome shock that occurs in allopolyploids [17]. The most accepted hypothesis for the association between allopolyploidy and apomixis, at least for gametophytic apomixis, is that apomixis can be caused by the asynchronous expression of duplicate genes [13, 14], which leads to important changes in the stages of the sexual program as early initiation of the embryonic sac and embryogenesis from atypical sites and times during the reproduction. According to the 'hybridization theory' postulated by Carman [14], the hybridization of two sexual ecotypes or related species with different reproductive characters contributes to the induction of apomictic phenomena. On a mechanistic level, this results from hybrid plants containing two sets of parental genes that are involved in female embryo sac development; the asynchronous expression of these duplicated genes leads to precocious embryo sac initiation and embryogenesis at aberrant sites during reproduction [18]. In this sense, the combination of hybridization and polyploidy would be responsible for changes in important processes during stages of megasporogenesis, megagametogenesis and fertilization, leading to apomixis [19].

Even though the frequent association between polyploidy/allopolyploidy and gametophytic apomixis is seen, several studies with tropical plants have demonstrated the occurrence of sporophytic apomixis (adventitious embryony) in polyploid taxa, which generally form polyploid complexes with a wide geographic distribution [20–22].

Several advantages are attributed to apomicts such as: (1) to allow reproduction even in the absence of pollinators; (2) to promote clonal reproduction through seeds, that is, to produce individuals genetically equal to the parental plant, but maintaining the advantages associated with the seeds as protection, dispersion and dormancy of the embryo; (3) to avoid costs associated with meiosis in cases where it is absent; (4) to avoid the costs associated with pollen grains' production in plants with low percentages of viable pollen grains or total male sterility; (5) to allow the establishment of extremely adapted genotypes to a specific environment, especially in apomictic species originated from polyploid hybrids, considered highly heterozygous and very vigorous [19]. For agriculture, apomixis is an attractive trait for the enhancement of crop species because it mediates the formation of large genetically uniform populations and perpetuates hybrid vigor through successive seed generations. Thus, apomixis is of practical interest to breeders as a means of fixation of highly complex genotypes including hybrid cultivars [13, 23]. Also, apomixis can be a viable alternative to increase the food production and, at the same time, reduce the deforestation of native vegetation [13].

2. Types of apomixis

Several types of apomixis have been identified in plants. In general, types of apomixis can be classified according to the origin and location of the somatic cell that gives rise to the embryo, the need or not of fecundation for the formation of endosperm and the possibility or not of occurrence along with sexual reproduction [18, 19]. The embryo can form within an embryo sac produced entirely from mitotic events in the gametophytic apomixis or directly from somatic cells surrounding the developing sexual embryo sac in the sporophytic apomixis. In some cases, fertilization is required for endosperm development and subsequent seed viability (pseudogamy), but in others, the endosperm can develop autonomously. Furthermore, most apomictic plants are facultative, meaning that sexual and asexual reproduction can occur concurrently [18].

In the gametophytic apomixis, changes in specific stages of the sexual process lead to the absence of reductive meiosis in the formation of the embryo sac (mitotic formation of unreduced embryo sacs), to the autonomous development of the embryo and adaptation in the formation of the endosperm. The embryo is formed parthenogenetically of the egg cell, located inside an unreduced embryo sac. This type of apomixis is subdivided into two forms, diplospory and apospory, depending on the origin of the cell that initiates unreduced embryo sac formation. In the diplospory, the unreduced embryo sac originates from the megaspore mother cell by suppression or modification of meiosis. In the apospory, the unreduced embryo sac originates from a somatic cell of the ovule, usually of the nucellus. Unlike diplospory in which there is commitment of the sexual reproduction, in the apospory, several nucellar cells can differentiate and give rise to multiple somatic embryonic sacs in a single ovule, while the megaspore mother cell originates a reduced embryo sac leading to the coexistence of the apomictic and sexual processes in the same ovule [12, 19].

In the sporophytic apomixis, also called adventitious embryony, the embryo originates directly from cells of the nucellus or the integument of the ovule. Commonly, adventitious embryony occurs associated with sexual reproduction and the apomictic embryo survival depends on successful fertilization of the meiotically derived embryo sac and its ability to grow sufficiently to gain access to the nutrient endosperm [12].

While adventitious embryony appears to be more common in tropical trees and shrubs, gametophytic apomixis is described primarily in herbaceous and perennial plants of temperate regions [24]. Gametophytic apomixis occurs most commonly among three families of angiosperms, Asteraceae, Rosaceae and Poaceae, but has been reported from at least 17 other families as well. For the plant families that occur in neotropical region, in addition to Asteraceae, Poaceae and Rosaceae, apospory was reported in Adoxaceae, Amaranthaceae, Araceae, Boraginaceae, Brassicaceae, Cucurbitaceae, Dioscoreaceae, Erythroxylaceae, Euphorbiaceae, Fabaceae, Hypericaceae, Lauraceae (Lindera), Magnoliaceae, Melastomataceae, Ochnaceae, Orchidaceae, Plantaginaceae, Polygonaceae, Ranunculaceae, Rutaceae and Urticaceae. Diplospory, also predominant in Asteraceae, Poaceae and Rosaceae, was recorded in other 16 families: Amaranthaceae, Amaryllidaceae, Balanophoraceae, Brassicaceae, Burmanniaceae, Cactaceae, Cucurbitaceae, Ochnaceae, Orchidaceae, Orobanchaceae, Plumbaginaceae, Potamogetonaceae, Rhamnaceae, Saururaceae, Thymelaeaceae and Urticaceae. Adventitious embryony, common in tropical vegetation, was registered in 48 plant families that occur in neotropics: Adoxaceae, Amaranthaceae, Amaryllidaceae, Anacardiaceae, Apiaceae, Apocynaceae, Araceae, Asparagaceae, Asteraceae, Bignoniaceae, Boraginaceae, Brassicaceae, Burseraceae, Cactaceae, Capparaceae, Caricaceae, Celastraceae, Clusiaceae, Combretaceae, Cucurbitaceae, Euphorbiaceae, Fabaceae, Lecythidaceae, Malpighiaceae, Malvaceae, Melastomataceae, Meliaceae, Moraceae, Myrtaceae, Ochnaceae, Onagraceae, Orchidaceae, Phyllanthaceae, Poaceae, Polygonaceae, Ranunculaceae, Rhamnaceae, Rosaceae, Rubiaceae, Rutaceae, Salicaceae, Sapindaceae, Smilacaceae, Solanaceae, Symplocaceae, Thymelaeaceae, Urticaceae and Zygophyllaceae. As previously mentioned, adventitious embryony was the most common type of apomixis in neotropical plants. In addition, this type of apomixis predominated in plant families with high diversity in this region. Examples are families well represented in the Brazilian vegetation such as: Apocynaceae (850 species in Brazil), Bignoniaceae (350 species), Euphorbiaceae (ca. 1000 species), Fabaceae (1500 species), Malvaceae (400 species), Myrtaceae (ca. 1000 species) and Orchidaceae (ca. 2500 species) [25]. Detailed information is in the next topic (apomixis in the neotropical region).

3. Apomixis in the neotropical region

Apomixis is relatively prevalent among angiosperms, with a pattern of distribution that suggests that it has evolved many times [13, 26]. It has been described in 32 orders, 78 families and 293 species [14, 26], and it is well represented among both monocotyledonous, eudicotiledonous and basal angiosperms [13]. Several authors have noted a marked bias in the distribution of apomixis among angiosperms. Of the plants known to use gametophytic apomixis, 75% of confirmed examples belong to three families, the Asteraceae, Rosaceae and Poaceae. Some authors have postulated that the current patterns of distribution may reflect the predisposition of certain plant groups to the unique developmental and genetic changes that characterize apomixis. But some of this bias also might relate to the ease of embryological examination in some plant groups or to data accumulated from embryological investigations associated with activities in crop improvement.

There are other noted associations between apomixis and various plant life history traits that provide insights into the nature and possible ecological role of this phenomenon. Apomixis frequently is associated with the expression of mechanisms that limit self-fertilization. Many apomictic plants belong to genera in which sexual members predominantly exhibit physiological self-incompatibility, dioecy or heterostyly [27]. An example in the neotropical region is Bignoniaceae. In this family, diploid taxa are generally self-incompatible, mainly exhibiting late-acting self-incompatibility [28]. However, polyploid species and populations often are self-compatible and exhibit asexual reproduction [22, 29].

Following are examples of orders and families well represented in the neotropical region that have apomictic taxa. Information on the occurrence of apomixis in families were extracted from Salomão and Allem [10], Carman [14], Batygina et al. ([30], Orchidaceae), Apomixis Database [31] and the open access articles. The occurrence of families in the neotropical region was confirmed in Flora do Brasil website [25], Angiosperm Phylogeny Group website [32], and Tropicos website [33].

3.1. Alismatales

Apomixis was reported in Araceae (*Aglaonema* and *Spathiphyllum*) and Potamogetonaceae (*Potamogeton*). In Araceae, adventitious embryony was determined for *Spathiphyllum patinii*, a Colombian species. In Potamogetonaceae, diplospory was observed in *Potamogeton obtusifolius*, species that does not occur in the tropical region.

3.2. Asparagales

Apomixis was reported in Amaryllidaceae, Asparagaceae, Hemerocallidaceae and Orchidaceae. In all families, gametophytic and sporophytic apomixis were recorded, but adventitious embryony predominate. In Orchidaceae, the most representative plant family in the neotropics among these cited families, polyembryony is frequent. The presence of several embryos in a single seed has been mentioned in certain tropical species, but this feature rarely occurs in species growing in temperate zone. Apomixis is present is 16 genera (*Cephalanthera, Corunastylis, Cynorchis, Cynosorchis, Dactylorhiza, Epidendrum, Epipactis, Gastrodia, Gymnadenia, Maxillaria, Neottia, Nigritella, Orchis, Spiranthes, Zeuxine* and *Zygopetalum*) [30]. Most genera of Orchidaceae species showed adventitious embryony, but *Corunastylis, Cynosorchis* and *Zygopetalum* showed both gametophytic and sporophytic apomixis. Only *Dactylorhiza* has aposporous taxa.

3.3. Asterales

In Asterales, apomixis was reported only in Asteraceae, the most diverse family of this order. In this plant family, gametophytic apomixis occurs in 27 genera and, adventitious embryony and diplospory were recorded in *Hieracium*. Apospory was recorded in *Calotis, Campovassouria, Campuloclinium, Cineraria* (*Senecio*), *Crepis, Eurybiopsis, Gutierrezia, Minuria* and *Pilosella*. Diplospory occurs in *Ageratina, Arnica, Blumea, Brachycome, Chondrilla, Chromolaena, Gyptis, Ixeris, Leontopodium, Praxelis, Rudbeckia, Taraxacum* and *Townsendia*. In genera such as *Antennaria, Erigeron, Eupatorium* and *Parthenium*, apospory and diplospory are observed. The phylogenetic distribution indicates that apomixis occurs mainly in the subfamily Asteroideae [34], restricting to polyploid cytotypes. Among tropical genera, the diplospory predominate.

3.4. Boraginales (?)

Apomixis was reported in three genera of Boraginaceae: *Cordia, Cynoglossum* and *Trichodesma*. Apospory is present in *Cordia* and *Cynoglossum*, while adventitious embryony is present in *Trichodesma*.

3.5. Brassicales

Gametophytic and sporophytic apomixis were reported in two genera of Brassicaceae: *Boechera* (apospory, diplospory and adventitious embryony) and *Iberis* (apospory and diplospory). Adventitious embryony was reported in Capparaceae (*Capparis*), Caricaceae (*Carica*) and Trapeolaceae (*Trapaeolum*).

3.6. Caryophyllales

Apomixis was reported in four families: Amaranthaceae, Cactaceae, Polygonaceae and Plumbaginaceae. In Amaranthaceae, diplospory was observed in the genus *Aerva*, while both gametophytic and sporophytic apomixis were observed in *Beta*. In Cactaceae is described the occurrence of adventitious embryony in *Haageocereus* (*H. tenuis*; [35]); and *Mammillaria, Opuntia* and *Pereskia*; in *Opuntia* was also reported the occurrence of diplospory. In Polygonaceae, apospory and adventitious embryony were reported in *Atraphaxis*, and in Plumbaginaceae, diplospory was observed in *Limonium* (*Statice*).

3.7. Celastrales

In Celastraceae is described the occurrence of adventitious embryony in *Celastrus* and *Euonymus*.

3.8. Cucurbitales

Apomixis was reported in five genera of Cucurbitaceae: *Cucumis* (apospory, diplospory and adventitious embryony), *Luffa* (diplospory), *Marah* (apospory), *Momordica* (diplospory and adventitious embryony) and *Sicyos* (adventitious embryony).

3.9. Dioscoreales

In this order, gametophytic apomixis was reported in two plant families: Burmanniaceae (*Burmmania* com diplospory) and Dioscoreaceae (*Tacca* with apospory).

3.10. Dipsacales

Apomixis was reported in Adoxaceae and Caprifoliaceae. In Adoxaceae, apospory and adventitious embryony were described in *Sambucus*, and in Caprifoliaceae, adventitious embryony was observed in the genus *Scabiosa*.

3.11. Ericales

In this order, apomixis was described in four families: Cyrillaceae, Lecythidaceae, Primulaceae and Symplocaceae. In Cyrillaceae, apospory was reported in *Cliftonia*. In the other families, adventitious embryony was registered in *Cariniana* (Lecythidaceae) (*C. estrellensis*, typical species of the neotropical savanna), *Ardisia* (Primulaceae) and *Symplocus* (Symplocaceae).

3.12. Fabales

Apomixis was reported only in Fabaceae, where there was a predominance of adventitious embryony. *Trifolium* was the only genus registered with apospory. *Acacia, Cassia, Copaifera* and *Millettia* showed adventitious embryony. *Acacia polyphila* and *Copaifera langsdorffii* are well-represented polyembryonic species of neotropical savannas [10].

3.13. Gentianales

In Gentianales, adventitious embryony was registered in Apocynaceae and Rubiaceae. In Apocynaceae is described the occurrence of adventitious embryony in *Cynanchum, Hancornia* and *Vincetoxycum*. *Hancornia speciosa* is a typical polyembryonic species of neotropical savannas [10]. In Rubiaceae, adventitious embryony was confirmed for *Genipa americana*; the type of apomixis was not determined in the genus *Coprosma*.

3.14. Lamiales

Apomixis was registered in three families: Bignoniaceae, Orobanchaceae and Plantaginaceae. In Bignoniaceae is described the occurrence of adventitious embryony in polyploid and polyembryonic species of *Anemopaegma* [21, 36, 37] and *Handroanthus* [22, 38] (**Figure 1**). In Orobanchaceae, diplospory was reported in *Cistanche* and *Orobanche*, and in Plantaginaceae, apospory was observed in *Globularia*.

3.15. Laurales

In this order, apospory was registered in Lauraceae in the genus *Lindera*.

Figure 1. Apomictic species of Bignoniaceae. (A–C) *Handroanthus chrysotrichus* (Mart. ex A. DC.) Mattos: (A) tree; (B) branches; (C) flowers. (D) *Anemopaegma arvense* (Vell.) Stellf. ex de Souza. (E) *Anemopaegma glaucum* Mart. ex DC. (F) Embryos of polyembryonic seeds of *Anemopaegma* species. (G and H) Polyembryonic seedlings of *Anemopaegma* species. Photos (A) and (B) provided by Rosana Farias-Singer; photo (C) provided by Rodrigo Bustus Singer; photos (D–H) provided by Fabiana Firetti.

3.16. Liliales

Apomixis was reported in Liliaceae, Melanthiaceae and Smilacaceae with the predominance of adventitious embryony. In Liliaceae, *Erythronium* and *Tulipa* (*Amana*) showed adventitious embryony, while in *Lilium* diplospory was observed. In Melanthiaceae, adventitious embryony was recorded in *Trillium*, and in Smilacaceae, it was recorded in *Smilacina*.

3.17. Magnoliales

Apomixis was reported in Annonaceae and Magnoliaceae. In Annonaceae, apomixis was suggested for the production of fruits in *Cymbopetalum brasiliense*, species from Atlantic forest; however, the type of apomixis was not determined [39]. In Magnoliaceae, apospory was recorded in *Woonyoungia*.

3.18. Malpighiales

Apomixis was recorded in eight plant families: Clusiaceae, Erythroxylaceae, Euphorbiaceae, Hypericaceae, Malpighiaceae, Ochnaceae, Phyllanthaceae and Salicaceae. Clusiaceae was characterized by adventitious embryony, which was recorded in *Clusia* and *Garcinia* [40]. In Erythroxylaceae, apospory was recorded in *Erythroxylum*. In Euphorbiaceae, adventitious embryony was recorded in *Alchornea* (*Coelebogyne*), *Cnidoscolus* (Jussieuia), *Euphorbia* and *Mallotus*, while apospory was observed in *Manihot*. In Hypericaceae, apospory was recorded in *Hypericum*. Malpighiaceae was characterized by adventitious embryony that occurred in all genera in which apomixis was recorded such as *Aspicarpa*, *Banisteriopsis*, *Byrsonima* (in *B. basiloba*, typical species of the *Cerrado* [10]), *Galphimia* (*Thryallis*), *Heteropterys*, *Hiptage*, *Malpighia* and *Stigmaphyllon*. However, occurrence of apospory was also observed in *Hiptage*. In Ochanacea, gametophytic and sporophytic apomixis were recorded in *Ochna*. In Phyllanthaceae, adventitious embryony was recorded in *Uapaca*, and in Salicaceae, this type of apomixis was observed in *Salix*.

3.19. Malvales

In this order, apomixis was reported in Malvaceae, Melastomataceae and Thymelaeaceae. In Malvaceae, particularly in the subfamily Bombacoideae, the occurrence of adventitious embryony was reported in *Bombacopsis* (*B. glabra*), *Pachira* (*P. oleaginea*) and *Eriotheca* (*E. gracilipes* and *E. pubescens* [20, 41]). The formation of polyembryonic seeds is reported in apomictic species of the group, as well as the occurrence of polyploidy. In Melastomataceae, most of the apomictic species occurs in Brazil. Apomictic species are described as autonomous, and in many of these, the total or partial unviability of the pollen grains is common [19, 51]. In this group, adventitious embryony was described in *Leandra* (*L. australis*), *Melastoma* (*M. malabathricum*), *Osbeckia* (*O. hispidissima*) and *Sonerila* (*Sonerila wallichii*); diplospory was recorded in *Leandra* (*L. australis*) and *Miconia* (*M. albicans*); and apospory was observed in *Clidemia* (*C. hirta*) and *Miconia* (*M. fallax*) [19]. In Thymelaeaceae, adventitious embryony was reported in *Daphne* and diplospory was registered in *Wikstroemia*.

3.20. Myrtales

This order was characterized by adventitious embryony that was recorded in Combretaceae, Myrtaceae and Onagraceae. In Combretaceae, apomixis was observed in *Combretum* (*Poivrea*). In Myrtaceae, adventitious embryony was recorded in *Callistemon, Eugenia* and *Syzygium*. In Onagraceae, adventitious embryony was reported in *Clarkia, Epilobium* (*Zauschneria*) and *Oenothera*.

3.21. Piperales

In this order, diplospory was recorded in Saururaceae in the genus *Houttuynia*.

3.22. Poales

Apomixis was recorded only in Poaceae that is characterized by apospory [14]. Apospory occurs exclusively in 41 genera (*Anthephora, Anthoxanthum, Apluda, Arundinella, Bouteloua, Brachiaria, Buchloe, Capillipedium, Chionachne, Chloris, Cortaderia, Dactylis, Danthonia, Echinochloa, Eragrostiella, Eremopogon, Eriochloa, Eustachys, Fingerhuthia, Harpochloa, Heteropogon, Hilaria, Hyparrhenia, Ischaemum, Lamprothyrsus, Merxmuellera, Panicum, Pentaschistis, Rendlia, Rhynchelytrum, Schima, Schizachyrium, Schmidtia, Setaria, Themeda, Thrasya, Thyrsostachys, Trachys, Tribolium, Tricholaena* and *Urochloa*). The occurrence of apospory with diplospory was recorded in *Dichanthium, Nardus* and *Paspalum*. Apospory and adventitious embryony were recorded in *Bothriochloa* and *Eulaliopsis*. Diplospory occurs exclusively in seven genera (*Calamagrostis, Elymus* (*Agropyron*), *Eragrostis, Phleum, Potamophila, Saccharum* and *Tripsacum*), and adventitious embryony occurs exclusively in *Bromus, Cenchrus* and *Oryza*. Apospory, diplospory and adventitious embryony were recorded in *Pennisetum, Poa* and *Sorghum*. In this family, apomixis is also related to polyploidy and is particularly common in tetraploids.

3.23. Ranunculales

Apomixis was registered in Ranunculaceae. Apospory was observed in *Ranunculus* and adventitious embryony was recorded in *Thalictrum*.

3.24. Rosales

In Rosales, apomixis was recorded in five families: Cannabaceae, Moraceae, Rhamnaceae, Rosaceae and Urticaceae. In Cannabaceae, apomixis was cited in *Humulus*, but the type of apomixis was not determined. In Moraceae, adventitious embryony predominate and was registered in *Ficus, Morus* and *Streblus*. In Rhamnaceae, diplospory was observed in *Pomaderris* and adventitious embryony was reported in *Zizyphus* (*Z. joazeiro*, species occurring in *Caatinga*, biome exclusive of Brazil). Rosaceae is characterized by gametophytic apomixis, which occurs exclusively in 10 genera (apospory in *Agrimonia, Amelanchier, Aphanes, Cotoneaster, Photinia* and *Sanguisorba*; apospory plus diplospory occur in *Crataegus, Rubus* and *Sorbopyrus*; diplospory occurs in *Parageum*). Despite the predominance of gametophytic apomixis in this family, adventitious embryony also occurs in several genera along with one or both types of gametophytic apomixis (adventitious embryony plus apospory in *Alchemilla, Chaenomeles, Malus* and *Pyrus*; adventitious embryony, apospory and diplospory in *Potentila* and *Sorbus*). In Urticaceae, diplospory occurs in *Boehmeria* and *Dorstenia* and gametophytic and sporophytic apomixis were registered in *Elatostema*.

3.25. Santalales

In this order, diplospory was described in Balanophoraceae, genus *Balanophora*.

3.26. Sapindales

In Sapindales, apomixis was reported in five families: Anacardiaceae, Burseraceae, Meliaceae, Rutaceae and Sapindaceae. In all families, adventitious embryony predominate, except in *Skimmia* (Rutaceae) where apospory occurs. In Anacardiaceae, adventitious embryony was registered in species with polyembryonic seeds of *Astronium, Mangifera, Myracrodruon* and *Tapirira; Astronium fraxinifolium, Myracrodruon urundeuva* and *Tapirira guianensis* are typical species of neotropical savannas of Brazil. In Burseraceae, adventitious embryony was registered in *Commiphora* (*C. lepthophloeos*, typical species of neotropical savannas of Brazil) and in *Garuga*. In Meliaceae, adventitious embryony occurs in *Aphanamixis* and *Lansium*. In Rutaceae, adventitious embryony also occurs in polyembryonic species of 10 genera (*Aegle, Citrus, Feroniella, Fortunella, Murraya, Poncirus, Ptelea, Ruta* (*Haplophyllum*), *Triphasia* and *Zanthoxylum* (*Xanthoxyllum*). Among the native representatives of Brazil, apomixis is described in *Galipea jasminiflora*. In Sapindaceae, adventitious embryony was described in polyembryonic species of *Magonia* (*M. pubescens*), which is also typical of neotropical savannas of Brazil.

3.27. Solanales

In this order, adventitious embryony was registered in five genera of Solanaceae: *Lycopersicon* (*Solanum*), *Petunia, Scopolia, Solanum* and *Withania*.

3.28. Zygophyllales

In this order, adventitious embryony was recorded only in Zygophyllaceae, genus *Zygophyllum*.

4. Apomixis and geographical distribution

In a number of plant and animal groups, it has been shown that sexual and asexual taxa exhibit different ranges of distribution, a phenomenon referred to as "geographical parthenogenesis" [42, 43]. Some studies conclude that parthenogenetic lineages tend to be found at higher latitudes, at higher elevations, in more xeric conditions, in more disturbed areas and on more island-like habitats than their sexual relatives. Bierzychudek [44] and Hörandl [45, 46] provided a comprehensive evaluation on the topic and concluded that apomictic plant groups have: (1) larger distributions, often considerably exceeding those of their sexual relatives; (2) tend to range higher latitudes and altitudes than their sexual relatives and (3) tend to colonize previously glaciated areas. It is also true that sexual relatives often have distributions centered within much larger ranges of apomictic complexes. Accordingly, geographical parthenogenesis seems to be applied only to taxa with gametophytic apomixis. Sporophytic apomixis, which is frequently observed in multiseeded tropical plants, has completely different developmental and control mechanisms and is connected with other ecological and geographical features [46]. However, Hojsgaard and Hörandl [47] demonstrated that sporophytic and gametophytic apomixis exhibit similar patterns of distribution across climate zones.

Taxa with uniparental reproduction, such as apomixis and self-fertilization, intrinsically possess the potential to colonize new environments by single individuals [46], because both breeding systems can provide reproductive assurance independently from pollinators or mating partners. In the case of apomixis, apomictic individuals do not require any pollinators or mating partners because they generate their seeds either independently of pollen (autonomous) or are able to use self-pollen for fertilization and formation of the endosperm (pseudogamy) [48]. Different hypothesis has been raised to explain the ability of colonization of marginal habitats by apomictic taxa [49]. However, these hypotheses focus on the fact that the great majority of apomictic taxa emerged from hybridization and polyploidization events, which confer to these taxa genomic features, which could provide more physiological and ecological flexibility to adapt to severe conditions. Such events are considered the main promoters of genomic evolution. The whole-genome doubling can be accompanied by gene silencing, diversification in gene expression, differential gene expression of gene copies, and epigenetic changes, important alterations for the generation of phenotypic plasticity and rapid adaptation [49]. In addition, the association among hybrid origin, polyploidization and sexuality in facultative apomixis establish multiple discrete arrays of "clones" featuring different and diverse genotypes (Vrijenhoek's frozen niche variation model). In this case, natural selection can cause partitioning among the "clonal" lineages and, thus, some apomictic genotypes may specialize on ecological niches outside the ecological optimum of the sexual parents [49].

Some examples of the neotropical flora corroborate the association among apomixis, wide geographic distribution and occupation of habitats with harsh conditions. In Bignoniaceae, some apomictic and polyploid species of *Anemopaegma* [21, 36, 37] and *Handroanthus* [22, 38] show wide geographical distribution and, although they occupy different vegetation types, they are characteristic of *Cerrado*, the neotropical savannas of central Brazil. The majority of apomictic species of *Anemopaegma* compose the *Anemopaegma arvense* complex (*A. acutifolium, A. arvense* and *A. glaucum*) that are typical of *Cerrado* areas that suffer constant disturbances, such as fire. As *Anemopaegma* species, *Handroanthus chrysotrichus* and *H. ochraceus* also form an agamic complex typical of the *Cerrado* areas.

Melastomataceae is a predominantly neotropical angiosperm family with many apomictic species [50, 51]. Most of the apomictic species of Melastomataceae belong to the tribe Miconieae and they have wider distributions than the sexual species of the same tribe [51, 52].

5. Genetic variability in apomicts

Traditionally, apomixis is considered a dead end of evolution due to the lack of genetic variability. This view was based on the assumption that loss of heterogeneity in populations would result in loss of potential to adapt to environmental changes [26, 53]. However, recent studies show that is not entirely true. The application of molecular markers in population studies has shown a significant genetic variability in natural apomictic populations, which can often be similar to that of populations of plants with sexual reproduction. Thus, the production of individuals genetically identical to the parental plant (i.e., clones in the strict sense) does not really exist in asexual reproduction [53].

According to Hörandl and Paun [53], different factors such as ancestry of apomictic lineages, backcrossing with sexual relatives, facultative sexuality and mutations contribute to high genetic diversity of apomicts. The ancestry of apomictic lineages defines the initial genetic diversity; thus, allopolyploid origin of most apomicts likely creates initially high levels of heterozygosity. The subsequent evolution of novel genotypes from the original lineages can occur via introgression, facultative sexuality and mutations. Introgression, backcrossing with sexual relatives, can be considered a source for the generation of new genotypes, but it depends on the level of sympatry among the asexual individuals and their sexual relatives. Most apomictic plants show facultative sexuality, which allows the formation of sexual and asexual offspring in the same individual and in the same generation. As a result of these processes of generation of genetic variability, natural apomict populations show greater genotypic variability than expected from a clonal population; however, the frequencies of such sexual events seem to be different among genera, microspecies and even populations of the same species. Finally, the accumulation of somatic mutations might also have a considerable impact on the levels of genetic variation in apomictic groups because of their gradual and continuous accumulation. Although most mutations in coding regions may have deleterious effects, the buffering effects of heterozygosity (allopolyploid origin) and the purging effects of facultative recombination (introgression plus facultative sexuality) have to be taken into account to reduce these effects.

In apomictic plants, genotypic diversity is partitioned more among than within populations [53]. However, studies of genetic diversity in populations of tropical apomictic plants, with polyembryonic seeds and supranumerary embryos formed by adventitious embryony, show a variation in this pattern and it seems to be species-specific. Larger variation was reported among populations of *Eriotheca gracilipes* (Bombacoideae-Malvaceae [41]) and *Garcinia mangostana* (Clusiaceae; [40]). However, in the *Anemopaegma arvense* species complex [54], *Clidemia hirta* [55], *Eriotheca pubescens* [41], *Handroanthus ochraceus* (Bignoniaceae; [38]) and *Opuntia ficus indica* (Cactaceae; [56]) were reported to have larger genetic diversity within populations. Apomictic populations of *Eriotheca* species presented values of genetic variability close (*E. pubescens*) or higher (*E. gracilipes*) than those of the conspecific sexual populations [41]. In *Handroanthus ochraceus*, genetic diversity found within the populations was also similar to that of allogamous populations of the species [38]. According to the authors, the dispersion of seeds by wind of the *Anemopaegma* species, *Eriotheca* species and *Handroanthus ochraceus* can lead to an increase in intrapopulational diversity, and it can also be responsible for the reduction of interpopulation differentiation due to the greater efficiency of the gene flow [38, 41, 54].

6. Apomixis as promoter of diversity

Apomixis seems to be taxonomically widespread in angiosperms with no clear tendency to specific groups [14, 26]. Despite the three large families Asteraceae, Poaceae and Rosaceae containing the majority of the known apomict-containing genera [26, 57], recent phylogenetic reconstructions of the major clades of angiosperms support the view of a broadly scattered distribution of apomixis over the entire phylogeny [58]. According to the survey carried out by Hojsgaard et al. [26], apomixis was registered in 32 orders (52%), 78 families (19%) and 293 genera (ca. 2.2%).

Some evidences like the presence of apomixis in highly diverse and cosmopolitan clades indicate that the apomixis can be considered a promoter of the diversification and evolution in angiosperms. As mentioned above, apomixis is usually closely related with hybridization and polyploidization events. Plant groups formed by these three processes are called agamic complexes. Such agamic complexes generally present wide distribution, which is related to the greater capacity of apomictic taxa of such complexes that occupy new ecological and geographical niches. It has been hypothesized that high levels of heterozygosity coupled to facultative sexuality and mutations increase genetic diversity and adaptive potential of apomict taxa. In addition, Carman [14] proposed a model for apomixis as facilitator of diversification through cycles of polyploidy and reversions to sex. According to this hypothesis, apomictic polyploids serve as pioneer explorers of new niches whereby they rapidly expand the distribution areas of their progenitor sexual populations by occupying new ecological and geographical niches. Thereafter, reversals to complete sexuality, accompanied by and possibly caused by substantial cytogenetic and genomic perturbations, allow for the establishment of new sexual populations in new habitats without the long-term disadvantages of asexuality [14, 26, 47, 58].

Studies on apomixis in tropical species are scarce and concentrated in few plant families such as Bignoniaceae, Malvaceae, Melastomataceae and Orchidaceae; however, there is a trend of the occurrence of apomict and polyploid taxa in more diverse clades of such neotropical families, as demonstrated by Hojsgaard et al. [26]. In Bignoniaceae, apomict and polyploid taxa were observed in *Anemopaegma* [21, 36] and *Handroanthus* [22, 38]. *Anemopaegma* includes 47 species of shrubs and lianas, representing the third largest genus of the tribe Bignonieae; polyploidy and apomixis were registered in shrub taxa that form the agamic complex called *Anemopaegma arvense* species complex [21, 29, 36, 59] and they show wide geographic distribution in neotropical savanna areas. In the genus *Handroanthus* with 67 species, polyploid and apomictic populations were observed in *H. chrysotrichus* and *H. ochraceus*; the first two species are interfertile and intermediate morphotypes are found in populations of both species. In Malvaceae, polyploidy associated with apomixis has been reported in *Eriotheca* species [20, 38]. Malvaceae is the most diverse family of the Malvales. Melastomataceae is the seventh largest family of flowering plants, with 170 genera and 5105 species of which some 3000 are neotropical; apomictic species of this family seems to be concentrated in the tribe Miconieae (88%). In Orchidaceae, apomixis is also associated with polyploidy and polyembryony and predominate in tropical species of this family [30].

7. Conclusions

Despite the predominance of sexual reproduction in tropical species, a growing number of studies have shown a reproductive mosaic in such species, with sexual and asexual reproduction in the same species in different populations. In predominant plant families of neotropical vegetation, apomixis, the asexual reproduction via seed, is closely related to polyploid taxa. In addition, adventitious embryony is the prevalent type of apomixis in several plant families that produce polyembryonic seeds and that show wide geographic distribution.

Although several studies have been conducted in recent years, knowledge on the distribution of apomixis in tropical taxa is still scarce and fragmentary. Most of the studies are concentrated in a few families such as Bignoniaceae, Malvaceae, Melastomataceae and Orchidaceae. Even in these more exploited families, only taxa of some clades had their reproductive system elucidated. Thus, I emphasize the need for further studies on the reproductive biology coupled with analyses of ploidy level in tropical species, especially at the population level. Studies concerning the genetic diversity of species that present a mosaic of reproductive systems should also be carried out along the geographical distribution of the species.

Author details

Fabiana Firetti

Address all correspondence to: ffiretti@gmail.com

Department of Botany, Institute of Biosciences, University of Sao Paulo, Brazil

References

[1] Thomas WW. Conservation and monographic research on the flora of Tropical America. Biodiversity and Conservation. 1999;**8**:1007-1015. DOI: 10.1023/A:1008857429787

[2] Schultz J. The Ecozones of the World: The Ecological Divisions of the Geosphere. Berlin: Springer; 2005. DOI: 9783540200147

[3] Antonelli A, Sanmartín I. Why are there so many plant species in the Neotropics? Taxon. 2011;**60**:403-414

[4] Forzza RC, Filardi FLR, Costa A, et al. (orgs.). Catálogo de Plantas e Fungos do Brasil. Vol. 1. Andrea Jakobsson Estúdio: Instituto de Pesquisa Jardim Botânico do Rio de Janeiro; 2010. ISBN: 978-85-88742-42-0

[5] Bawa KS. Breeding systems of tree species of a lowland tropical community. Evolution. 1974;**28**:85-92. DOI: 10.1111/j.1558-5646.1974.tb00729.x

[6] Bawa KS, Perry DR, Beach JH. Reproductive biology of tropical lowland rain forest trees. I. Sexual systems and incompatibility mechanisms. American Journal of Botany. 1985; **72**:331-345

[7] Oliveira PE, Gibbs EP. Reproductive biology of woody plants in a cerrado community of Central Brazil. Flora. 2000;**195**:311-329. DOI: 0367-2530/00/195/04-311

[8] Machado IC, Lopes AV, Sazima M. Plant sexual systems and a review of the breeding system studies in Caatinga, a Brazilian tropical dry forest. Annals of Botany. 2006;**97**:277-287. DOI: 10.1093/aob/mcj029

[9] Jaimes I, Ramírez N. Breeding systems in a secondary deciduous forest in Venezuela: The importance of life form, habitat and pollination specificity. Plant Systematics and Evolution. 1999;**215**:23-36. DOI: https://doi.org/10.1007/BF00984645

[10] Salomão AN, Allem AC. Polyembryony in angiospermous trees of the Brazilian cerrado and caatinga vegetation. Acta Botanica Brasilica. 2001;**15**:369-378. DOI: http://dx.doi.org/10.1590/S0102-33062001000300007

[11] Koltunow AM. Apomixis: Embryos sacs and embryos formed without meiosis or fertilization in ovules. The Plant Cell. 1993;**5**:1425-1437. DOI: 10.1105/tpc.5.10.1425

[12] Koltunow AM, Grossniklaus U. Apomixis: A developmental perspective. Annual Review of Plant Biology. 2003;**54**:547-574. DOI: 10.1105/tpc.150710

[13] Bicknell RA, Koltunow AM. Understanding apomixis: Recent advances and remaining conundrums. The Plant Cell. 2004;**16**:S228-S245. DOI: https://doi.org/10.1105/tpc.017921

[14] Carman JG. Asynchronous expression of duplicate genes in angiosperms may cause apomixis, bispory, tetraspory, and polyembryony. Biological Journal of the Linnean Society. 1997;**61**:51-94. DOI: 10.1111/j.1095-8312.1997.tb01778.x

[15] Grimanelli D. Epigenetic regulation of reproductive development and the emergence of apomixis in angiosperms. Current Opinion in Plant Biology. 2012;**15**:57-62. DOI: 10.1016/j.pbi.2011.10.002

[16] Mizarghaderi G, Hörandl E. The evolution of meiotic sex and its alternatives. Proceedings of the Royal Society B: Biological Sciences. 2016;**283**:1-10. DOI: 10.1098/rspb.2016.1221

[17] Chen ZJ. Genetic and epigenetic mechanisms for gene expression and phenotypic variation in polyploid plants. Annual Review of Plant Biology. 2007;**58**:377-406. DOI: 10.1146/annurev.arplant.58.032806.103835

[18] Tucker MR, Koltonow AMG. Sexual and asexual (apomictic) seed development in flowering plants: Molecular, morphological and evolutionary relationships. Functional Plant Biology. 2009;**36**:490-504. DOI: 10.1105/tpc.011742

[19] Caetano AP de S, Cortez PA. Reprodução assexuada. In: Rech AR, Agostini K, Oliveira PE, Machado IC, editors. Biologia da Polinização. Editora Projeto Cultural; 2014. pp. 93-111

[20] Mendes-Rodrigues C, Carmo-Oliveira R, Talavera S, Arista M, Ortiz PL, Oliveira PE. Polyembryony and apomixis in *Eriotheca pubescens* (Malvaceae-Bombacoideae). Plant Biology. 2005;**7**:533-540. DOI: 10.1055/s-2005-865852

[21] Firetti-Leggieri F, Lohmann LG, Alcantara S, Costa IR, Semir J. Polyploidy and polyembryony of *Anemopaegma* (Bignonieae, Bignoniaceae). Plant Reproduction. 2013;**26**:43-53. DOI: 10.1007/s00497-012-0206-3

[22] Bittencourt Jr NS, Moraes CIG. Self-fertility and polyembryony in South American yellow trumpet trees (*Handroanthus chrysotrichus* and *H. ochraceous*, Bignoniaceae): A histological study of postpollination events. Plant Systematics and Evolution. 2010;**288**:59-76. DOI: https://doi.org/10.1007/s00606-010-0313-2

[23] Galla G, Zenone S, Avesani L, Altschmied L, Rizzo P, Sharbel T, Barcaccia G. Pistil transcriptome analysis to disclose genes and gene products related to aposporous apomixis in *Hypericum perforatum* L. Frontiers in Plant Science. 2017;**8**:79. DOI: 10.3389/fpls.2017.00079

[24] Whitton J, Sears CJ, Baack EJ, Otto SP. The dynamic nature of apomixis in the angiosperms. International Journal of Plant Science. 2008;**169**:169-182. DOI: 10.1086/523369

[25] Flora do Brasil 2020 in construction. Jardim Botânico do Rio de Janeiro. Available from: http://floradobrasil.jbrj.gov.br [Accessed 2017-07-20]

[26] Hojsgaard D, Klatt S, Baier R, Carman JG, Hörandl E. Taxonomy and biogeography of apomixis in angiosperms and associated biodiversity characteristics. Critical Reviews in Plant Sciences. 2014;**33**:414-427. DOI: http://dx.doi.org/10.1080/07352689.2014.898488

[27] Asker SE, Jerling L. Apomixis in Plants. Boca Raton: CRC Press; 1992

[28] Gibbs PE. Late-acting self-incompatibility—The pariah breeding system in flowering plants. New Phytologist. 2014;**203**:717-734. DOI: 10.1111/nph.12874

[29] Firetti-Leggieri F. Biossistemática das espécies do complexo *Anemopaegma arvense* (Vell.) Stellf. ex de Souza (Bignoniaceae, Bignonieae): Aspectos anatômicos, citológicos, moleculares, morfológicos e reprodutivos [thesis]. Campinas: Universidade Estadual de Campinas; 2009

[30] Batygina TB, Bragina EA, Vasilyeva VE. The reproductive system and germination in orchids. Acta Biologica Cracoviensia Series Botanica. 2003;**45**:21-34. DOI: 10.1.1.899.7010

[31] Apomixis Database. Department of Systematics, Biodiversity and Evolution of Plants. Georg-August-Universität Göttingen. Available from: http://www.apomixis.uni-goettingen.de; [Accessed: 2017-09-11]

[32] Angiosperm Phylogeny Group Website. Available from: http://www.mobot.org/MOBOT/research/APweb/ [Accessed: 2017-08-27]

[33] Tropicos. Missouri Botanical Garden. Available from: http://www.tropicos.org/ [Accessed: 2017-08-26]

[34] Noyes RD. Apomixis is Asteraceae: Diamonds in the Rough. Functional Plant Science and. Biotechnology. 2007;**1**:207-222

[35] Arakaki M, Speranza P, Soltis PS, Soltis DE. Genetic variability of an unusual apomictic triploid cactus—*Haageocereus tenuis* Ritter—from coast of Central Peru. Journal of Heredity. 2013;**104**:127-133. DOI: 10.1093/jhered/ess072

[36] Firetti-Leggieri F, Costa IR, Forni-Martins ER, Lohmann LG, Semir J. Chromosome studies in Bignonieae (Bignoniaceae): The first records of polyploidy in *Anemopaegma* Mart. ex Meisn. Cytologia. 2011;**76**:185-191

[37] Sampaio DS, Bittencourt Jr NS, Oliveira PE. Sporophytic apomixis in polyploid *Anemo paegma* species (Bignoniaceae) from Central Brazil. Botanical Journal of the Linnean Society. 2013;**173**:77-91. DOI: 10.1111/boj.12076

[38] Mendes MG. Variabilidade genética e morfológica em populações de *Handroanthus ochraceus* (Bignoniaceae) com sistemas reprodutivos e ploidias distintos [thesis]. Uberlândia: Universidade de Uberlândia; 2013

[39] Braun M, Dötterl S, Gottsberger G. Absence of pollinators and apomictic fruit production in an Atlantic rainforest population of *Cymbopetalum brasiliense* (Annonaceae). Plant Systematics and Evolution. 2011;**296**:265-273. DOI: 10.1007/s00606-011-0493-4

[40] Ramage CM, Sando L, Peace CP, Carroll BJ, Drew RA. Genetic diversity revealed in the apomictic fruit species *Garcinia mangostana* L. (mangosteen). Euphytica. 2004;**136**:1-10. DOI: https://doi.org/10.1023/B:EUPH.0000019456.06040.eb

[41] Marinho RC. Variabilidade genética e variação de ploidia em espécies de *Eriotheca* (Bombacoideae-Malvaceae) com diferentes sistemas reprodutivos [thesis]. Universidade Federal de Uberlândia; 2013

[42] Vendel A. La parthénogenese géographique. Contribution a' l'etude biologique et cytogique de la parthénogenese naturelle. Bulletin Biologique de la France et de la Belgique. 1928;**62**:164-281

[43] Molins MP, Corral JM, Aliyu OM, Koch MA, Betzin A, Maron JL, Sharbel TF. Biogeographic variation in genetic variability, apomixis expression and ploidy of St. John's wort (*Hypericum perforatum*) across its native and introduced range. Annals of Botany. 2014;**113**:417-427. DOI: 10.1093/aob/mct268

[44] Bierzichudek P. Patterns in plant parthenogenesis. Experientia. 1985;**41**:1255-1264. DOI: https://doi.org/10.1007/BF01952068

[45] Hörandl E. The complex causality of geographic parthenogenesis. New Phytologist. 2006;**171**:525-538. DOI: 10.1111/j.1469-8137.2006.01769.x

[46] Hörandl E. Evolutionary implications of self-compatibility and reproductive fitness in the apomictic *Ranunculus auricomus* polyploid complex (Ranunculaceae). International Journal of Plant Science. 2008;**169**:1219-1228. DOI: 10.1086/591980

[47] Hojsgaard D, Hörandl E. Apomixis as a facilitator of range expansion and diversification in plants. In: Pontarotti P, editor. Evolutionary Biology: Biodiversification from Genotype to Phenotype. Springer International Publishing Switzerland; 2015. pp. 305-327

[48] Hörandl E. The evolution of self-fertility in apomictic plants. Sexual Plant Reproduction. 2010;**23**:73-86. DOI: 10.1007/s00497-009-0122-3

[49] Schinkel CCF, Kirchheimer B, Dellinger AS, Klatt S, Winkler M, Dullinger S, Hörandl E. Correlations of polyploidy and apomixis with elevation and associated environmental gradients in an alpine plant. AoB Plants. 2016;**8**:plw064. DOI: 10.1093/aobpla/plw064

[50] Goldenberg R, Shepherd GJ. Studies on the reproductive biology of Melastomataceae in 'cerrado' vegetation. Plant Systematics and Evolution. 1998;**211**:13-29. DOI: https://doi.org/10.1007/BF00984909

[51] Santos APM, Fracasso CM, Santos ML, Romero R, Sazima M, Oliveira PE. Reproductive biology and species geographical distribution in the Melastomataceae: A survey based on New World taxa. Annals of Botany. 2012;**110**:667-679. DOI: 10.1093/aob/mcs125

[52] Goldenberg R. Apomixia como alternativa à reprodução sexuada em Melastomataceae. In: Cavalcanti TB, Walter BMT, editors. Tópicos atuais de Botânica. Brasília: EMBRAPA—Recursos Genéticos; 2000. pp. 225-230

[53] Hörandl E, Paun O. Patterns and sources of genetic diversity in apomictic plants: Implications for evolutionary potentials. In: Hörandl E, Grossniklaus U, Van Djik P, Sharbel T, editors. Apomixis: Evolution, Mechanisms and Perspectives. Lichtenstein: ARG Gantner Verlag KG; 2007. pp. 169-194

[54] Batistini AP. Diversidade morfológica, genética e química de populações naturais de *Anemopaegma arvense* (Vell.) Stellf. [thesis]. Universidade Estadual Júlio de Mesquita Filho; 2006

[55] DeWalt SJ, Hamrick JL. Genetic variation of introduced Hawaiian and native Costa Rican populations of an invasive tropical shrub, *Clidemia hirta* (Melastomataceae). American Journal of Botany. 2004;**91**:1155-1162. DOI: 10.3732/ajb.91.8.1155

[56] Bendhifi M, Baraket G, Zourgui L, Souid S, Salhi-Hannachi A. Assessment of genetic diversity of Tunisian Barbary fig (*Opuntia ficus indica*) cultivars by RAPD markers and morphological traits. Scientia Horticulturae. 2013;**158**:1-7. DOI: http://dx.doi.org/10.1016/j.scienta.2013.04.015

[57] Richards AJ. Plant Breeding Systems. UK: Chapman & Hall; 1997. 529 p

[58] Hörandl E, Hojsgaard D. The evolution of apomixis in angiosperms: A reappraisal. Plant Biosystematics. 2012;**146**:681-693. DOI: http://dx.doi.org/10.1080/11263504.2012.716795

[59] Firetti F, Zuntini AR, Gaiarsa JW, Oliveira RS, Lohmann LG, Van Sluys M-A. Complete chloroplast genome sequences contribute to species delimitation - A case study of the Anemopaegma arvense species complex. American Journal of Botany. 2017;**104**:1493-1509. DOI: https://doi.org/10.3732/ajb.1700302